AF784698

CONTENTS

REVIEW OF THE GENERAL ACCOUNTING OFFICE REPORT ON CARGO PREFERENCE REQUIREMENTS

HEARING

BEFORE THE

SUBCOMMITTEE ON FOREIGN AGRICULTURE AND HUNGER

OF THE

COMMITTEE ON AGRICULTURE HOUSE OF REPRESENTATIVES

ONE HUNDRED THIRD CONGRESS

SECOND SESSION

SEPTEMBER 29, 1994

Serial No. 103–89

Printed for the use of the Committee on Agriculture

U.S. GOVERNMENT PRINTING OFFICE

87–784 WASHINGTON : 1995

For sale by the U.S. Government Printing Office
Superintendent of Documents, Congressional Sales Office, Washington, DC 20402

ISBN 0-16-046807-8

COMMITTEE ON AGRICULTURE

E (KIKA) DE LA GARZA, Texas, *Chairman*

GEORGE E. BROWN, JR., California, *Vice Chairman*
CHARLIE ROSE, North Carolina
DAN GLICKMAN, Kansas
CHARLES W. STENHOLM, Texas
HAROLD L. VOLKMER, Missouri
TIMOTHY J. PENNY, Minnesota
TIM JOHNSON, South Dakota
BILL SARPALIUS, Texas
JILL L. LONG, Indiana
GARY A. CONDIT, California
COLLIN C. PETERSON, Minnesota
CALVIN M. DOOLEY, California
EVA M. CLAYTON, North Carolina
DAVID MINGE, Minnesota
EARL F. HILLIARD, Alabama
JAY INSLEE, Washington
THOMAS J. BARLOW III, Kentucky
EARL POMEROY, North Dakota
TIM HOLDEN, Pennsylvania
CYNTHIA A. MCKINNEY, Georgia
SCOTTY BAESLER, Kentucky
KAREN L. THURMAN, Florida
SANFORD D. BISHOP, JR., Georgia
BENNIE G. THOMPSON, Mississippi
SAM FARR, California
PAT WILLIAMS, Montana
BLANCHE M. LAMBERT, Arkansas

PAT ROBERTS, Kansas, *Ranking Minority Member*
BILL EMERSON, Missouri
STEVE GUNDERSON, Wisconsin
TOM LEWIS, Florida
ROBERT F. (BOB) SMITH, Oregon
LARRY COMBEST, Texas
WAYNE ALLARD, Colorado
BILL BARRETT, Nebraska
JIM NUSSLE, Iowa
JOHN A. BOEHNER, Ohio
THOMAS W. EWING, Illinois
JOHN T. DOOLITTLE, California
JACK KINGSTON, Georgia
BOB GOODLATTE, Virginia
JAY DICKEY, Arkansas
RICHARD W. POMBO, California
CHARLES T. CANADY, Florida
NICK SMITH, Michigan
TERRY EVERETT, Alabama
FRANK D. LUCAS, Oklahoma
RON LEWIS, Kentucky

PROFESSIONAL STAFF

DIANNE POWELL, *Staff Director*
VERNIE HUBERT, *Chief Counsel and Legislative Director*
GARY R. MITCHELL, *Minority Staff Director*
JAMES A. DAVIS, *Press Secretary*

SUBCOMMITTEE ON FOREIGN AGRICULTURE AND HUNGER

TIMOTHY J. PENNY, Minnesota, *Chairman*

CHARLIE ROSE, North Carolina, *Vice Chairman*
THOMAS J. BARLOW III, Kentucky
CYNTHIA A. MCKINNEY, Georgia
SCOTTY BAESLER, Kentucky
CHARLES W. STENHOLM, Texas

WAYNE ALLARD, Colorado
TOM LEWIS, Florida
JOHN T. DOOLITTLE, California
CHARLES T. CANADY, Florida
TERRY EVERETT, Alabama

REVIEW OF THE GENERAL ACCOUNTING OFFICE REPORT ON CARGO PREFERENCE REQUIREMENTS

THURSDAY, SEPTEMBER 29, 1994

HOUSE OF REPRESENTATIVES,
SUBCOMMITTEE ON FOREIGN AGRICULTURE AND HUNGER,
COMMITTEE ON AGRICULTURE,
Washington, DC.

The subcommittee met, pursuant to call, at 9:30 a.m., in room 1302, Longworth House Office Building, Hon. Timothy J. Penny (chairman of the subcommittee) presiding.

Present: Representatives Rose, Barlow, Allard, Doolittle and Canady.

Also present: Representative Condit, member of the committee.

Staff present: Dale Moore, minority legislative coordinator; Brenda Connolly, clerk; Jane Shey, Bruce White, Anita R. Brown, James A. Davis, and Lynn Gallagher.

OPENING STATEMENT OF HON. TIMOTHY J. PENNY, A REPRESENTATIVE IN CONGRESS FROM THE STATE OF MINNESOTA

Mr. PENNY. The hearing will come to order.

This is the second hearing our subcommittee has held on the issue of cargo preference. The first hearing covered the broad issues of cargo preference, and, as part of the review, I requested GAO to report on cargo preference requirements in food aid shipments.

At a time when food aid needs in Rwanda, the Horn of Africa and elsewhere challenge our ability to address the problem of global hunger, cargo preference requirements in food aid merit closer examination.

At the same time Congress has begun consideration of reforms to U.S. maritime and agricultural policy, we are marking the 40th anniversary of Public Law 480, the primary law authorizing international food assistance. In anticipation of next year's farm bill, I thought it all the more appropriate for us to begin a focus on this particular question.

In a time of severe budget constraints, we have to take a very close look at cargo preference requirements, which obviously add to the cost of delivering food aid, and which may not be serving even the stated purposes of the cargo preference law itself.

I am always interested in ensuring the efficiency of Government programs; this is particularly important when programs are designed to help meet the needs of the world's hungry.

With that as an opening remark, I would indicate that our purpose this morning is simply to receive testimony from the GAO regarding their most recent study.

I appreciate the fact that Senator Breaux and others have taken an interest in this issue, and so we trust that there may be some action on this question on the Senate side of the Hill at some point. But, for now, our goal is to get into the subcommittee record the work of the General Accounting Office, and our hope is that, after this session, the next Congress will pick up where we have left off and follow through on the recommendations that flow from this report.

As always, it is good to have you here, Mr. Mendelowitz. I will call on you and allow you to divide the testimony as you wish among your colleagues. We always appreciate the thoroughness of the work that you do in your office and look forward to your testimony this morning.

STATEMENT OF ALLAN I. MENDELOWITZ, MANAGING DIRECTOR, INTERNATIONAL TRADE, FINANCE, AND COMPETITIVENESS, U.S. GENERAL ACCOUNTING OFFICE, ACCOMPANIED BY PHILLIP J. THOMAS, ASSISTANT DIRECTOR; SCOTT EINHORN, SENIOR EVALUATOR; SUSAN HOFFMAN, SENIOR EVALUATOR, NEW YORK REGIONAL OFFICE; AND SHEILA RATZENBERGER, SENIOR LEGAL ADVISOR, OFFICE OF GENERAL COUNSEL

Mr. MENDELOWITZ. Thank you, Mr. Chairman.

In recent years, we have testified before you a number of times on research undertaken by the General Accounting Office at your request, and we have always greatly appreciated the good use to which you put our work. And because it seems that this will be the last time we will have the opportunity to testify before you because of your impending retirement, we just wanted to express that appreciation and wish you well.

I am accompanied today by principal GAO staff who assisted in this project.

On my far right is Phil Thomas, the Assistant Director responsible for our work on international agriculture issues; on my immediate right, Susan Hoffman, who is the evaluator in charge of our review of the application of cargo preference to food aid programs; on my immediate left, Scott Einhorn, who was the assignment manager on this project. And to Mr. Einhorn's left is Sheila Ratzenberger, our Assistant General Counsel who worked with us on this project.

With your permission, I would like to make a few summary comments and submit my full statement for the record.

Mr. PENNY. Without objection.

Mr. MENDELOWITZ. The objectives of cargo preference are to maintain U.S.-flag ships to serve as a naval and military auxiliary in time of war or national emergency and to carry a substantial portion of U.S. waterborne domestic and international commerce. In pursuit of these objectives, 75 percent of food aid cargo are reserved for U.S.-flag ships to the extent they are available at fair and reasonable rates.

At your request, we reviewed the application of cargo preference to food aid programs to determine, first, the extent to which the legislative objectives of cargo preference are being realized; second, how cargo preference requirements affect U.S. food aid programs; and, third, how USDA and AID management of food aid transportation affects the cost of transportation in the program.

In our review we have concluded, with respect to the first objective, that the application of cargo preference requirements to food aid programs does not significantly contribute to helping achieve the objectives of the Merchant Marine Act of 1936. The U.S.-flag ships that are most dependent on food aid preference cargoes are not currently viewed by the Department of Defense as militarily useful.

In addition, food aid preference cargoes do not significantly contribute to ensuring that U.S.-flag ships carry a substantial portion of either U.S. domestic or foreign waterborne commerce.

The U.S.-flag ships that the Department of Defense currently views as militarily useful, those that provide liner service, are either supported by Jones Act trade—that is, domestic waterborne cargo, ie. cargo carried between two U.S. ports—or largely dependent upon the maritime industry's operating differential subsidies. These subsidies enable the liners to compete for foreign commercial cargoes because they offset their operating costs to the extent they are greater than their foreign competitors.

Furthermore, the application of cargo preference under current rules discourages U.S. shipowners from taking steps to reduce the costs of operating ships under cargo preference. Investing in new ships is discouraged because of the 3-year waiting period currently imposed on cargo preference eligibility for foreign-built U.S.-flag ships and it is possible to successfully operate inefficient ships in the food aid cargo preference trade.

Thus, applying cargo preference laws to food aid programs makes it possible for U.S. shipowners to maintain inefficient and commercially noncompetitive U.S.-flag ships that do not significantly contribute to the ability of the U.S. merchant marine to carry commerce other than food aid and which increases the cost to transport U.S. food aid preference cargo.

In addition, the higher cost of using U.S.-flag ships and their limited availability can adversely affect the operation of the U.S. food aid programs by, first, reducing the amount of commodity that can be purchased and, second, prohibiting a country from purchasing the lowest price or the variety of commodity desired. This situation occurs because commodity purchasing decisions are driven more by the availability of U.S.-flag ships than the availability of commodities.

Food aid transportation costs are also affected by USDA and AID management practices. USDA and AID have been using ocean transportation contract terms as a means to provide additional services to recipient countries. Shipowners must comply with these terms even though they may result in higher shipping costs.

We believe that some potential may exist for savings in food aid transportation costs by removing the uncertainty associated with estimating the cost of these services by adopting terms more consistent with those used for similar commercial cargos.

Food aid transportation costs are also increased because the majority of food aid shipments occur in the last half of the calendar year, instead of being more evenly spaced throughout the year. While there is no conclusive reason why this happens, it is clear that food aid transportation costs could be reduced if food aid shipments were more evenly spaced throughout the year and demand for the limited number of U.S. flag vessels was not concentrated into a short period of time.

In our report, we make a number of recommendations and provide some matters for consideration to the Congress.

With respect to the matters for consideration for Congress, if Congress continues to support the objectives of cargo preference and is willing to continue to devote resources to that end, Congress may wish to consider a more efficient alternative for achieving those objectives.

For example, a program like the current operating subsidy program which will expire in 1998 could be used to support those ships and their crews that DOD finds militarily useful and that could also successfully compete for U.S. foreign commercial cargos.

If Congress decides to continue to apply cargo preference to food aid programs, it may wish to consider giving U.S. shipowners incentives to invest in more modern and efficient ships in order to reduce food aid transportation costs such as allowing new, foreign-built U.S.-flag ships to immediately carry food aid preference cargos.

Regarding our recommendations, if Congress wishes to continue the application of cargo preference to food aid programs and acts to permit U.S.-flag foreign-built ships to immediately carry food aid preference cargo, we have several recommendations to the Department of Transportation and the Department of Agriculture and AID which we believe could reduce the cost of transporting U.S. food aid under the cargo preference laws, such as:

In the case of the Department of Transportation, instead of providing a guideline rate for each individual ship based on its actual operating cost, we suggest that an average for comparable size ships be used so that those who have the most efficient ships have an added incentive to bid and take advantage of the available business.

With respect to the Secretary of Agriculture and the Administrator for the Agency for International Development, we have some suggestions that deal with experimenting with contract terms for the transportation of food aid cargos that are more consistent with contract terms used for commercial cargos and trying to encourage recipient countries to more evenly space their food aid shipments throughout the year.

Mr. Chairman, this concludes my summary comments, and we will be happy to try to answer questions you and your committee members may have.

[The prepared statement of Mr. Mendelowitz appears at the conclusion of the hearing.]

Mr. PENNY. I thank you for your testimony. I appreciate the succinct manner in which you summarized the report.

I note on the very first paragraph of your conclusions that you make the point that U.S. food aid preference cargos do not significantly contribute to ensuring that U.S.-flag ships carry a substantial portion of either U.S. domestic or foreign waterborne commerce. Do you have some statistical material in the report to share with us this morning that backs up that statement?

Mr. MENDELOWITZ. Approximately 4 percent of all U.S. waterborne cargo is Government-impelled cargo. So cargo preference in all its applications only accounts for about 4 percent of U.S. waterborne international commerce. The cargo preference application to the U.S. food aid programs result in less than 1 percent of all U.S. waterborne international cargos being carried on U.S.-flag vessels.

So we have, first of all, a very small share of U.S. international or foreign commerce being carried on U.S.-flag ships as a result of application of cargo preference to the food aid programs; and, second, because these ships are such high-cost ships in terms of their operating costs and their lack of efficiencies, the ships that carry the bulk of U.S. food aid cargos under cargo preference are not able to compete for any other kind of cargo and therefore, in effect, are primarily used to carry food aid cargo.

Mr. PENNY. It is an exclusive situation in which they basically exist for this program. Is that fair to say?

Mr. MENDELOWITZ. Basically. If you look at the overall statistics, about 84 percent of the food aid cargo is carried by a relatively small number of bulk cargo ships, 16 or 18 a year, plus a few tug-barge combinations. Once in a while, a tanker with operating differential subsidy jumps into the food aid cargo to pick up a cargo here and there.

But, basically, the 84 percent carried by the bulk cargo ships represents the only cargo that these ships are able to carry and compete for. The other 16 percent of the food aid cargo is shipped on liners. And for at least three-quarters of the liners that carry some of the cargo preference food aid cargoes, the food aid shipments are not the reason why, they told us, they maintained their U.S. flag status. For about 25 percent of the liners that do carry some food aid preference cargo, it represents a more important share of their cargo.

So that, if you sort of work backwards, the 84 percent of the cargo going to the bulk carriers is carried by ships that don't carry anything else. The remaining 16 percent is carried by liners—this tends to be high-value products, packaged types of products, not bulk commodities like wheat or other types of grain. For most of the liners that carry this business, the preference cargo represents such a small part of what they carry that it is not really that important to them. And for a small subset of liners, about 25 percent of those liner ships that carry cargo preference, it is of some importance.

Mr. PENNY. In some respects, it is comparing apples to oranges since these liners would be carrying finished products, packaged products, as opposed to the bulk commodities on the others.

Mr. MENDELOWITZ. Exactly.

Mr. PENNY. But the subsidy we pay, is there any way that we can sort of even out the differences and make some comparisons about the level of subsidy we pay to these ships that are not so dependent on the cargo preference policy, as opposed to those ships that are totally dependent or nearly totally dependent?

Mr. MENDELOWITZ. I think it is important to sort of keep in mind the different kinds of ocean freight services provided and the different kinds of ocean freight vessels.

A charter ship is a ship which basically is used when you have a large amount of goods to ship. You charter the ship to go from point "A" to point "B," and you basically determine what port it is going to load at, what it is going to load, where it is going to go and where it is going to off-load. And, typically, this is the type of cargo service offered by bulk carriers.

Liners, on the other hand, provide service on a fixed route. A liner typically keeps to a scheduled route serving particular ports on a particular schedule. And the type of technology used for carrying goods in liner traffic is typically what is called containerized or intermodal shipment. Large containers, 20-foot or 40-foot-long containers are filled with cargo and carried from inland sites to the ports on either trucks or trains and then lifted up, put on the ships, taken to the destination port, off-loaded and then trucked to wherever they have to go.

The majority of the food provided under food aid programs by tonnage is, in fact, bulk commodities. It is the 84 percent that I was talking about that goes under cargo preference on bulk carriers.

The way these particular ships receive their operating subsidies, in effect, is they have this 75 percent of the food aid cargo business reserved for U.S.-flag ships. The maximum rates that they may receive are set based on their operating costs. And when bidding to carry this cargo is opened up, U.S.-flag carriers bid and compete for the 75 percent only against other U.S.-flag ships. And the rates they receive are obviously much higher than foreign-flag rates because they reflect the fact that U.S.-flag ships have greater operating costs.

So these ships receive their subsidy in the form of higher rates to carry the bulk commodities for the particular shipments that they are chartered.

Liners, on the other hand, receive a subsidy from the Federal Government, in effect, on an annual basis. If they ply a route on a set schedule, they receive a lump sum of money to help cover operating costs. They then go out and compete for business with shipping fees that are competitive with foreign-flag rates. When they receive a payment for carrying preference cargo, it is, in fact, a payment equal to their regular shipping tariff. The subsidy they receive comes in the form of the general subsidy received to maintain their operations throughout the year.

So for the program, the subsidy comes in different ways depending upon the type of service being provided and the type of ship.

Mr. PENNY. What do we know about the shipping firms that are so reliant on cargo preference shipments? Are there relatively few firms involved? Are there a lot of firms with just a narrow sliver of all the ships involved in this program or are there firms that are totally reliant on this program and their entire fleet basically tied to Public Law 480 shipments?

Ms. HOFFMAN. There are all kinds of different situations for these companies. We went as part of our work and we interviewed a lot of the companies that carry food aid, and you will find all different types of situations, some that are very heavily reliant on these cargoes, some that are not, some that it is just a piece of their business. So it really depends upon the nature of the company.

Mr. PENNY. So there is no rule of thumb here.

Ms. HOFFMAN. No, there isn't.

Mr. PENNY. When we talk about these ships being almost exclusively used for Public Law 480 shipments, that doesn't necessarily mean that it is an entire company with a fleet of ships totally reliant on the program.

Ms. HOFFMAN. Not necessarily.

Mr. PENNY. In most cases, is it fair to say that it is maybe one portion of a company's total—what would we call these? Total fleet?

Ms. HOFFMAN. In most instances, yes; that is true.

Mr. PENNY. I note your reference to the Jones Act. Could you just for the record spell out for this subcommittee the provisions of the Jones Act and is that primarily tied to domestic vessels, for domestic shipment?

Mr. MENDELOWITZ. The Merchant Marine Act of 1920, which is commonly referred to as the Jones Act, reserves all domestic waterborne cargo for U.S.-flag ships. Any waterborne commerce between two U.S. ports, such as between the west coast of the United States and Alaska, between the west coast of the United States and Hawaii, between any two ports on the east coast or between the east coast and the gulf coast, are reserved for U.S.-flag ships by the Jones Act.

There is no Government outlay in support of these particular ships because, basically, only U.S.-flag ships can carry this type of cargo. U.S.-flag ships are available and in fact compete for the business at whatever the market will bear.

Mr. PENNY. The last paragraph—well, I will see. The first complete paragraph on page 22, you make reference to food aid transportation costs are also increased because a majority of food aid shipments occur in the last half of the calendar year instead of being more evenly spaced throughout the year. While there is no conclusive reason why this happens, it is clear that food aid transportation costs could be reduced if food aid shipments were more evenly spaced throughout the year. I realize your caveat that there is no conclusive reason available as to why this happens, but could you speculate or elaborate as to some of the factors that surfaced during your review?

Mr. MENDELOWITZ. One of the reasons is it takes some time at the beginning of each fiscal year to enter into new agreements with recipient countries so that if the first half of the year is the time spent negotiating new agreements, the last half of the year is the time when the commodities are, in fact, shipped.

Second, many of these recipient countries have limited storage capacity, limited port facilities, and they need the commodities in certain times of the year and so that tends to contribute to bunching of the cargos in the last part of the year.

Mr. PENNY. Realistically, is there any way to spread this out throughout the year?

Mr. MENDELOWITZ. I don't think that it is possible to get an even distribution with one-twelfth of all cargo shipped in each month. But I think that it is possible to get a more even distribution, and I think we should look for creative ways to do it.

For example, one possibility is if you could reach an agreement in one year, and permit some shipments in the following year under that agreement, that would provide a way of providing additional shipments in the first 6 months of the next year.

Mr. PENNY. Anyone else want to add a suggestion along those lines?

[No response.]

Mr. PENNY. You also mentioned that we might more effectively achieve the stated goals of the cargo preference law if we identified those ships that were actually of use to the military and focus our subsidies on shipments involving those ships. That would effectively cut out bulk ships configured for bulk commodities, wouldn't it?

Mr. MENDELOWITZ. If you look at what has happened over the past 25 years with respect to the military needs, objectives, and shipping technology, a lot has changed. With respect to shipping technology, there has been a very substantial shift away from break bulk cartage to containerized intermodal shipping.

If you look at military needs, their needs have changed along with the changing transportation technology. Just as private sector commercial cargos are shifting to containerized or intermodal shipment, the Department of Defense is also in the midst of increasing its use of containers. Therefore, it is looking for container ships in order to be able to efficiently carry containers.

And if you look at military cargos themselves, the nature of military cargos has changed. Military equipment has gotten bigger. Military equipment has gotten heavier. This large, heavy equipment is best carried on "RoRo" ships—roll on, roll off—which enable the tank carriers and various other equipment to drive right onto the ship and drive off in the port of destination.

And it is this changing technology, changing military needs, changing nature of military equipment which have all contributed to the fact that ships such as modified bulk carriers which may have been useful 25 years ago are no longer considered particularly useful. So that I think that your observation that we probably wouldn't be supporting bulk ships under the cargo preference law is probably correct.

Mr. PENNY. You mentioned that 16 percent of our cargo preference shipments are on ocean liners, which would fall into the category of container-type vessels. Is there a way that we could modify cargo preference so that we would essentially move away from subsidizing the bulk cargo, which obviously would be cheaper if we didn't use U.S.-flag vessels? But is there a way that we could subsidize even more container shipments under Public Law 480 so that the number of those vessels would be increased and we would then have more of those under U.S.-flag for military purposes in times of an emergency?

Mr. MENDELOWITZ. The Congress has very clearly stated in the Maritime Act of 1936 that the objectives of cargo preference are to create a naval and military auxiliary in time of war or national emergency and to carry a substantial portion of U.S. waterborne cargo.

I think the issue is to focus on what the legislative objectives are, and then in light of changed circumstances assess how those objectives can most efficiently and effectively be achieved in light of circumstances that exist today.

When we looked at the application of cargo preference for food aid, given that most of the commodities being carried are bulk commodities which require bulk cargo ships, it is hard for us to see how the objectives of the Merchant Marine Act of 1936 can be efficiently and effectively achieved through the application of cargo preference to the food aid program.

And that is why we suggested the matter for the consideration of the Congress, which is if Congress continues to support these important objectives, given the fact that cargo preference is in fact funded out of budgetary expenditures, if the Congress wishes to continue to devote resources to these important objectives, the Congress should think in terms of how to achieve the objectives in the most efficient and effective way possible.

Given that it would be hard to modify cargo preference to efficiently and effectively achieve the objectives of the Merchant Marine Act of 1936, we suggested that Congress may wish to consider alternative ways of achieving those objectives, namely providing a type of operating differential subsidy to the specific kinds of ships that the military finds useful. At the same time, because the subsidy would in effect offset the higher operating costs of U.S.-flag vessels, such a subsidy would permit U.S. crewed, U.S.-flag liners to compete for commercial cargo.

Mr. PENNY. I see. I am going to ask the same question in a different way. Of the 25 percent of our food aid shipments that are not subject to cargo preference, have you run the statistics to determine how much of that volume is containerized and how much is bulk? Is there sort of a typical pattern over the years? Is it comparable to the same ratios under the cargo preference law, 84–16?

Mr. MENDELOWITZ. Yes. The impression we are under is it is basically about the same.

Mr. PENNY. Basically about the same?

Mr. MENDELOWITZ. Yes.

Mr. PENNY. So, again, it speaks to the fact that there isn't a terribly effective way to put a greater emphasis on these containerized vessels through this particular law.

Mr. MENDELOWITZ. I think that is correct. The essential problem is that bulk commodities like wheat, soybeans, and corn are commodities that require bulk ships to be efficiently transported.

Mr. PENNY. I understand that. And the balance, the 25 percent of these shipments that are not brought under the subsidy program, are reflective of the 75 percent that are brought under.

Mr. MENDELOWITZ. That is correct.

Mr. PENNY. The amount of that would be bulk cargo is pretty much consistent with the 84 percent of the shipments that are bulk cargo under the cargo preference shipments.

Mr. MENDELOWITZ. Yes.

Mr. PENNY. In the second to the last page of your testimony, you recommend the Secretary of Transportation should instruct the Administrator of the Maritime Administration to promote the efficiency of the ships that carry food aid preference cargos. One way this can be done is by changing the method of calculating guideline rates so that average operating costs for all similar size ships, instead of actual operating costs for each individual ship, are considered. Could you spell out in more detail what you are getting at with that recommendation and how would it make a difference?

Mr. MENDELOWITZ. This particular recommendation is linked to an earlier recommendation or matter for consideration, and that is the issue of what U.S.-flag ships are permitted to carry cargo preference cargos.

Under current rules, a U.S.-flag ship, if it was built in a foreign shipyard, must operate for 3 consecutive years as a U.S.-flag ship before it is eligible to carry cargo preference cargo. Now, given that U.S.-flag ships have much higher operating costs—there are higher crew costs, there are a number of other costs associated with U.S.-flag ships that other flag ships don't have—it is virtually impossible for a U.S.-flag carrier to take a foreign-built ship, reflag it as a U.S. ship, and operate it for 3 years in order to qualify to carry preference cargo.

So, in effect, the current system makes it almost impossible to get a newer, more efficient ship into the cargo preference trade. So that the only way the change in setting guideline rates could contribute to improving the efficiency of the program is if it, in fact, were possible under the program to get newer, more efficient ships into the U.S. flag trade to carry cargo preference.

Now, if a foreign-built ship that is reflagged as a U.S. ship and meets all U.S. flag requirements and regulations established by the Coast Guard—if a foreign-built U.S.-flag ship was immediately permitted to start carrying cargo preference, then this change in the computation of the guideline rate becomes important.

As currently established, the Maritime Administration creates a guideline rate for each individual ship based upon the actual operating cost of that ship. It, in effect, is a cost-plus computation.

Now, if it turns out that demand for U.S.-flag ships to carry preference cargo is low during a certain point in time that bids are solicited, U.S. ships may not necessarily receive the guideline rate. Ships bidding their maximum permitted rate may not secure cargo if there is a relatively small volume of cargo and the ships that are available far exceed the capacity needed to carry that cargo.

But when you bunch all of the cargo under Public Law 480 and the other food aid programs into a relatively short period of time, you create a relatively large demand for a relatively small number of ships. Competitive bidding doesn't drive down the price. Inefficient ships that may be laid up realize they can successfully bid to carry some cargo.

And in that situation what happens is each ship that carries preference cargo receives its guideline rate, or its cost-plus rate.

Now, if the maximum rate were set based on an average of all the ships in a particular class that are operating, it means that the maximum rate that you could get would be set by an average of the actual cost of operating the ships. Inefficient ships would be unable to recover their full costs, and the most efficient ships would in fact be able to make a little extra profit if in fact they are able to get the maximum average guideline rate.

And, in that situation, you would provide incentives for additional new ships to be brought in. And as new ships get brought in that are more efficient, the average maximum rate for that class of ship would in fact decline over time, improving the efficiency of the U.S. fleet and reducing the cost of cargo preference.

Mr. PENNY. Thank you for that clarification. I appreciate that. Mr. Barlow.

Mr. BARLOW. Thank you, Mr. Chairman.

Again, I want to compliment the General Accounting Office for the tremendous work that they do and the very thorough work that they do on all the issues that you all investigate for us. We would be hard put if we didn't have you as a resource, let me just say that.

This is a very fine report, very thorough. And I am particularly interested in the recommendations you make at the end of your testimony. I think that they are on point in a number of areas. I would hope, looking ahead, that we are able to combine some of the legislative measures we pass in Congress this year for new ship construction in the United States and the exploration of new marine technology in those construction programs with the cargo preference.

I was struck, for example, just seeing a picture recently in a magazine of some of the propellers that the German shipyards are turning out. They are amazing. They are double and triple propellers, and they are curved in very different ways, nothing that I have seen the United States come up with.

And I believe both the West Europeans and the Japanese are way ahead of us in ship design. I hope that we can use cargo preference and the construction programs of MARAD to explore some of this new technology.

But what I was particularly interested with, on the point that you and the Chairman have been discussing, about the bulk carriers and the fact that we have gone into—in the international marine trade, as all of us know—we have gone into the cargo container technology and the fact that military technology has changed, too, and that they can use these cargo containers much more efficiently, say, than the bulk carriers.

And here is the question. Wouldn't you say that we will be in situations we have been in the past few years in situations around the world—where there is a disaster or relief is needed after a military upheaval, such as Africa or possibly other parts of the world. Africa's been in the news in the last few years—where you don't necessarily need the fast load-on, load-off capabilities, the high-tech military capabilities that the Pentagon would need from the newest technology, but you need the—for want of a better phrase—the old-time bulk carriers—efficient, low-cost, get the bulk over there to Africa, to an African port. Wouldn't there be a place for these ships in some future scenario that the State Department might cook up?

Mr. MENDELOWITZ. I think that there are a number of considerations that would have to be looked at in the scenario that you have laid out.

One, of course, would be the type of cargo that has to be carried. Typically, emergency relief cargo would tend to be flour rather than wheat, if you are trying to avert hunger. Flour is a commodity that is shipped in bags, typically it is not a bulk commodity, and it would be something that a break bulk carrier would carry.

Secondly, we would have to look at the nature of the ports. A lot of these countries that we are talking about are countries which have minimal port facilities. And if there were this kind of emergency, you probably would want to have a ship which can carry the type of commodity that is used to avert hunger, such as bag flour, and that would be self-sustaining, such as a break bulk carrier with sufficient on-board winches and cranes to load and off-load the pallets. This, quite honestly, is a technology for transportation that is extremely inefficient by today's standards.

If the Congress and the administration felt that it was important to have some ships like this in the fleet for the types of emergency that you described, then I think that it is certainly possible to do. But I think it would have to be an explicit decision that we want these ships for these particular situations.

Mr. BARLOW. No. Those are very good points. And in crisis situations very possibly you are right. But not all of them. I can see situations coming up where you would be going into a port that has the bulk facilities for off-loading and yet your zone of contention is maybe two or three countries away. But that is the closest port.

Mr. MENDELOWITZ. Sure.

Mr. BARLOW. And you get it in there, and they may have the facilities to break it down and repackage it.

Mr. MENDELOWITZ. Right.

Mr. BARLOW. And there are situations, wouldn't you say, of an international nature where you are a nation rebuilding. That might go on for years. Or there is a population explosion that is putting tremendous pressure on a healthy economy or healthy political structure, but they just need some long-term help. And we do have situations like that now where we are shipping there, and there are facilities there, and the bulk carriers are the cheapest way to get it there. Wouldn't you say that is true?

Mr. MENDELOWITZ. If you are carrying bulk commodities like wheat, as I indicated, the most efficient way to carry them is by bulk carrier.

The question, though, is whether in fact you must have U.S.-flag bulk carriers to carry that. The situation in which that it is most difficult to get a foreign-flag ship to carry cargo in pursuit of U.S. objectives is when there, in fact, is a threat to the ship and its crew because of hostile action.

If a ship is going into a war zone, we can rely on American-flag ships to do the right thing under our existing laws. If we rely on a foreign-flag to carry cargo into a war zone, we always run the risk that, as happened one or two times in the Desert Storm–Desert Shield operation, a couple crews may jump ship.

But in the situation that you described, as long as there isn't a war situation, and there isn't threat of hostile action against the ship and the crew, I don't know that we would have difficulty finding foreign-flag bulk carriers that would be willing to carry the bulk cargos to the destinations you have described—and do it quite efficiently.

Mr. BARLOW. Well, it is not just the availability of foreign-flag carriers. These ships do serve as a training ground for merchant marine seamen who are moving up the ladder to a more technologically advanced shipping structure, such as the rapid off-load ships. There is that point, too, isn't there?

Mr. MENDELOWITZ. I think you have raised an excellent point. When you look at the ability of the U.S.-flag fleet to serve as a naval auxiliary in time of war or emergency, you need both ships and crews. And there is no question that the civilian crews that are skilled mariners that currently man the U.S.-flag fleet are an important part of this naval auxiliary.

Mr. BARLOW. They don't jump ship in the Persian Gulf.

Mr. MENDELOWITZ. No. One, they don't jump ship; and, two, the point I was going to make was that the skills of a skilled mariner are not something that are learned overnight. Handling cargo and sailing ships is a very complex and technically demanding undertaking that requires a tremendous amount of training to do it well, and to do it safely. So the U.S. able-bodied, skilled mariners who man the U.S.-flag fleet are an important part of the ability of the Navy to activate its Ready Reserve Fleet in the event of an emergency.

If you look at the current Ready Reserve Fleet and you look at the current pool of U.S.-skilled mariners, it turns out that each berth of a U.S.-flag cargo ship supports about 2.1 to 2.2 skilled mariners because, basically, seamen work about half the year. When you put to sea, you put to sea for a long stretch, and you don't get your weekends off. You don't get holidays off. You work 7 days a week. You work around the clock, in effect, because you are at sea. And whether you are in your bunk or not, you are there if something happens.

And so, typically, you don't find mariners working 12 months out of the year if they want to have any kind of family lives. What we have is a Ready Reserve Fleet that requires maybe 3,500 to 4,000 skilled mariners, and currently there is a large pool, relative to that need, of U.S.-skilled mariners who are available to man the Ready Reserve Fleet.

If you look at the number of mariners, for example, who operate the bulk ships carrying cargo preference under the food aid programs, you are dealing with a relatively small number of seamen who, if those ships weren't U.S.-flag ships, wouldn't make—given current numbers of available mariners a major difference in the availability of skilled mariners.

Down the road, though, if the U.S.-flag fleet continues to contract, this might be a problem. And I think that the Maritime Administration and Department of Defense have to think very carefully about the most efficient way of maintaining the necessary level of skilled mariners.

One of the problems, for example, that was pointed out to me was that one of the reasons why there was difficulty in finding in a few instances some skilled mariners for ships activated from the

87-784 95-2

U.S.-Ready Reserve Fleet is the fact that a mariner who makes himself available to man the Ready Reserve Fleet is treated differently than a member of the armed services who is in the Reserves and mobilized in the event of an emergency.

A member of the Army Reserve, the Marine Reserve or the Naval Reserve has reemployment rights. If the Government calls up a member of the Reserve to meet a national emergency, when that member is demobilized from active service that member of the Reserves is guaranteed reemployment at his former job. A skilled mariner who volunteers to man the Ready Reserve Fleet during an emergency doesn't have those same reemployment rights.

Those are the types of things that we have to look at to ensure that, you have the skilled mariners available when needed.

Mr. PENNY. Have you looked at that?

Mr. MENDELOWITZ. The scope of this report was not a broad assessment of all the issues involved in manning the Ready Reserve Fleet. There is other GAO work underway on that.

But I would say that, based on what I have learned about the needs of the Ready Reserve Fleet and the availablity of manpower, I personally think it would be a good idea to put skilled mariners on the same legal footing with respect to reemployment rights as members of the uniformed services Reserve.

Mr. BARLOW. Those fellows who jumped ship in the Persian Gulf, jumped off of foreign ships?

Mr. MENDELOWITZ. That is correct.

Mr. BARLOW. And they were foreign crews?

Mr. MENDELOWITZ. I don't know exactly, but I remember being told there were two foreign flag ships whose foreign crews while at port in the Mediterranean jumped ship.

Mr. BARLOW. That is because we didn't have sufficient capacity under American flag, is that right? That is why we had to hire those ships and crews.

Ms. HOFFMAN. I just wanted to add, at that point in Desert Shield-Desert Storm the military sealift command contracted for all those ships. They didn't go out and require shipowners to use their ships in that particular situation. They contract both foreign and U.S. flag ships to suit their purposes. Plus, there were a lot of ships donated to the cause.

Mr. MENDELOWITZ. If you look at the mobilization of ships during Desert Storm-Desert Shield, what happened was there was an immediate surge need. The immediate surge need was filled by foreign-flag contract ships. And once we had a chance to get the Ready Reserve Fleet going, the U.S.-flag fleet was able to take over.

So if you look at the numbers of U.S.-flag versus foreign-flag ships that took part in carrying cargo under Desert Shield-Desert Storm, there were a large number of foreign-flag ships and a relatively small number of U.S.-flag ships.

But if you look at the share of cargo carried, if I remember the numbers correctly, the U.S. flag ships carried the vast majority of all the cargo. And the reason why is that the foreign-flag ships were typically chartered for one trip when the surge need was there. We were just getting started. Once the U.S.-flag ships were available, they basically were available for the whole period of time and went back and forth picking up, carrying, and discharging cargo.

Mr. BARLOW. I will just sum up. I think you have an excellent report here. You have some very good recommendations about policy changes to the U.S. fleet. And I am assuming you made some of these recommendations on the basis of the legislation that just passed Congress, which calls for shipbuilding programs in the United States for U.S. flag ships. And I am hoping that we will see some new technologies come along, and I hope that those new technologies are tested out in the cargo preference trade.

Thank you very much. Excellent testimony.

Mr. PENNY. Mr. Rose.

Mr. ROSE. Thank you, Mr. Chairman.

I am sorry I haven't heard all of the direct testimony by the GAO, but this is not a new issue for me or for the GAO. And I think my friend, Mr. Mendelowitz, knows my position on this.

Under the cargo preference law, 75 percent of the U.S. Government-financed food aid shipments must be transported on U.S.-flag vessels if available at fair and reasonable rates.

Mr. Chairman, I ask unanimous consent that my statement be made part of the record. And so I will skip around here for a minute to hit just a few high points.

[The prepared statement of Mr. Rose follows:]

OPENING STATEMENT OF CONGRESSMAN CHARLIE ROSE

Under the cargo preference laws, 75 percent of U.S. government-financed food aid shipments must be transported on U.S.-flag vessels, if available at fair and reasonable rates. Bulk food aid shipments are carried by bulk carriers, tankers, and tug/barge units. Individualized parcels are carried by liner vessels. The crux of the GAO report is its assumption that the principal reason for food aid cargo preference is to provide DOD with vessels it can charter for military shipments during wars and national emergencies and its conclusion that the program does not meet that objective, perhaps because DOD chartered only liner vessels during Operation Desert Shield/Desert Storm.

GAO misses the point. First, bulk carriers, tankers, and tug/barge units are militarily useful. Operation Desert Shield/Desert Storm was unique because there was a long mobilization period, virtually the entire world was united against Iraq, and Saudi Arabia provided DOD with perhaps the world's most modern deep-water, containerized port, combined with a virtually limitless source of inexpensive oil and state-of-the-art desalinization facilities. The next conflict may require the use of bulk carriers, tankers, and tug/barge units to carry grain, water, oil and other products, as well as the liner vessels that carry individual parcels. The definition of "militarily useful" can change radically in a different conflict under different conditions, where there could be vessel losses, such as in the Iran/Iraq war, that could raise the value of any type of vessel rather than excluding all but the most ideal militarily useful vessels.

Second, food aid cargo preference has several other objectives, in addition to providing militarily useful vessels for DOD charter. There is a strong foreign policy justification, which has widespread American public support, that the recipients of our foreign aid ought to see it arrive on ships flying the American flag. How visible and how effective is the American commitment if our aid arrives on flag-of-convenience vessels?

Third, the GAO study does not recognize that food aid cargo preference is one of several programs supporting the U.S.-flag merchant marine, and that the vessels and crews involved in food aid transportation have an important national security role. Cargo preference ensures an adequate base of U.S.-flag vessels available for shipping not only grain and oil to our armed forces, but also food aid to our friends and allies, some of which may be subject to a boycott by foreign-flag operators.

Fourth, as DOD but not GAO recognizes, the crewmembers aboard food aid vessels are readily available to work aboard Ready Reserve Fleet (RRF) and other vessels used by DOD during wars and national emergencies. It would cost DOD untold millions of dollars to maintain this capability if the U.S. lost this manpower base -- a fact not considered by GAO.

Fifth, the very existence of cargo preference reflects the longstanding congressional recognition that it is more expensive to operate as a U.S.-flag carrier than as a foreign-flag-of-convenience carrier because of an array of U.S. government-imposed tax, safety, environmental, health-benefit, and labor requirements. Federal and state income taxes, social security, Medicare, and unemployment alone double labor costs. Cargo preference is simply a decision by the U.S. government that it will help offset the cost of U.S.-flag operation by reserving a portion of U.S. government-generated cargoes to U.S.-flag vessels, if available at fair and reasonable rates.

Cargo preference for food aid should be viewed in a broader context. Food aid programs, because of the Buy-American and Ship-American requirements, create jobs and income here in America. They help American farmers, food processors, maritime, and many other industries. American commodities and American shipping may cost more because they have to compete with subsidized foreign competition and/or foreign competition virtually exempt from tax and regulatory requirements. But that is no reason to buy foreign commodities and use foreign shipping. Do the American people really want to use their tax dollars to throw American farmers and seamen out of work for the benefit of foreigners? GAO, of course, cannot consider these broader issues.

The one area where GAO is correct is its recommendations that USDA and AID adopt the commercial practices, to the extent feasible, that will result in lower ocean transportation costs. According to USDA, the average rate for foreign-flag bulk carriers is $39 per ton and the average rate for U.S.-flag carriers is $69 per ton. GAO found that Israel charters U.S.-flag bulk carriers in the low $30 per ton range, significantly less than the foreign-flag rates paid by USDA, because Israel contracts on a long-term basis and uses commercial terms. The Israeli experience indicates the huge potential savings resulting from USDA's adoption of commercial-like practices.

Mr. ROSE. Bulk carriers, tankers, and tug barge units are militarily useful. Camp LeJeune and Fort Bragg are part of my congressional district, and Operation Desert Storm-Desert Shield was unique because there was a long mobilization period. Virtually the entire world was united against Iraq. Saudi Arabia provided DOD with perhaps the world's most modern deepwater container port combined with a virtually limitless source of inexpensive oil and state-of-the-art desalinization facilities.

The next conflict may require the use of tankers, tug-barge units to carry grain, water, oil and other products, as well as the liner vessels. The definition of militarily useful can change radically in a different conflict under different conditions where there could be vessel losses, such as the Iran-Iraq war, that could raise the value of any type of vessel rather than excluding all but most ideally militarily useful vessels. Do you disagree with that?

Mr. MENDELOWITZ. We asked the Department of Defense specifically on a number of occasions how they viewed the ships that basically are U.S.-flag ships and operating on the high seas because of cargo preference applied to food aid.

We went ship by ship discussing the characteristics, capabilities, and flexibility of these ships. We asked the Department of Defense how they would use these ships and when they would use these ships.

And the Department of Defense was very clear that these particular ships—and it is a relatively small number of ships that are maintained at relatively high cost under cargo preference—these particular ships are, in fact, a very last resort for DOD. They basically said they would not use these ships if there was anything else that was available. They were quite clear.

Mr. ROSE. What does anything else available mean? Foreign flag?

Mr. MENDELOWITZ. It could be foreign flag. It could be——

Mr. ROSE. They said that? They said they would rather have foreign-flag ships if they were available than the U.S. merchant marine fleet? I find that hard to believe.

Mr. MENDELOWITZ. For example, we asked about the capability of the bulk carriers in the preference cargo trade which had been modified to carry containers, which is the preferred form of transportation for much of the DOD cargo. We went through a list of some 16 or 18 of these bulk carriers, which are the ships that were maintained at an annual average cost of over the 1991–1993 cargo preference years, in round numbers, of $200 million.

And they said, basically, the entire container carrying capacity of this fleet of ships could be equaled by two modern container ships that DOD could buy for about $100 million. These two modern container ships would have the ability to carry as much as and load much more quickly the cargo carrying capacity of the entire 16 or 18 ships. DOD was quite explicit in responding to our question. And, we probed on this point quite deeply, Mr. Rose, because we wanted to make sure that we hadn't somehow missed some subtlety with respect to this issue. And, basically, they said that this is just not an efficient means of providing for their required ocean cargo carrying capacity.

Mr. ROSE. So DOD has pretty well predicted that most conflicts in the future will only be in countries where they have containerized modern ports.

Mr. MENDELOWITZ. Mr. Rose, I would be overstepping my cognizance if I were to substitute my judgment for the Department of Defense on national security issues. All I can do on this issue is report back to you what the Department of Defense told us quite emphatically and quite clearly with respect to these ships.

Mr. ROSE. Did you put in the record who you talked to at the Department of Defense?

Mr. MENDELOWITZ. We did by command and by agency within the Department of Defense. I don't think we listed the officials by name.

Mr. ROSE. I notice on page 77 of the report you talk about shipping rates for grain under the Israeli side letter agreement are significantly lower than food aid to aid shipments to Russia. Well, why don't we make long-term agreements with American-flag ships to have that kind of lower rates, stability of the rate over a longer period of time?

Mr. MENDELOWITZ. I think that is an excellent point. If the volumes of food aid cargo to individual destinations were sufficiently large to provide for a long term, multivoyage contracts, that would be possible. But for most of the destinations of food aid cargo, the volumes involved are not sufficient to provide for these multiple voyage long-term contracts.

Mr. ROSE. I haven't read the whole report. I believe you just released it yesterday or today?

Mr. MENDELOWITZ. Today.

Mr. ROSE. People complain about us voting on bills sometimes we haven't read. Sometimes we don't get to read the report before we hear about it. But you do, of course, recognize in here the cost of—the little added—the difference of shipping to Russia versus shipping to other places.

Mr. MENDELOWITZ. Yes.

Mr. ROSE. And such things as you are not sure who is picking up your cargo or whether the Russian Mafia gets all of it.

Mr. MENDELOWITZ. I think you have a good feel for the problems of doing business in Russia.

Mr. ROSE. And you don't find that dealing with American-flag ships in an instance of shipping to Russia might be highly preferential to your local Baltic bottom?

Ms. HOFFMAN. We, if you look at appendix II in the report, it talks about the "Russian Food for Progress Program," which we looked at——

Mr. ROSE. If you will get closer to the microphone. Thank you.

Ms. HOFFMAN. Appendix II of the report talks in more detail about the "Russian Food for Progress Program" this last year, about what happened with those particular cargos. And when we looked at that report——

Mr. ROSE. Summarize. What did happen to some of those cargos?

Ms. HOFFMAN. For those cargoes for that specific program there really were not any significant delays at the port, and the U.S. shipowners didn't have any significant problems.

Mr. ROSE. These were U.S. ships.

Ms. HOFFMAN. Yes.

Mr. ROSE. OK.

Ms. HOFFMAN. And prior shipments that weren't part of this particular program, when we talked to the shipowners they expressed that they did have some difficulties.

Mr. ROSE. These were non-U.S. ships.

Ms. HOFFMAN. No, U.S.-flag ships. But in the most recent——

Mr. ROSE. What is the general reputation of the problem shipping to Russia?

Ms. HOFFMAN. Well, what we found in this most current program to Russia, there really weren't——

Mr. ROSE. I am not talking about the current programs. I am talking about if you are going to send your mother's heirloom furniture to your new house in St. Petersburg, Russia, are you going to feel real comfortable that it is going to get there? And, if so, which shipping company are you going to prefer to get it there? That is all I am asking.

Based on what you know from studying and writing this report you certainly developed some impression and feeling about shipping things to Russia, haven't you?

Ms. HOFFMAN. But I only looked at those associated with food and so I don't think I could really answer the question.

Mr. ROSE. OK, it is tea cakes you are sending. You are going to have a big party at your new house in St. Petersburg. Talk to me about that. You really think there is no particular benefit from shipping all on American bottom. Ma'am?

Ms. HOFFMAN. Not that I could see from the work that we did. It really didn't make much of a difference for the food aid programs with respect to the ability of the food to get to its destination, whether it was on a U.S.-flag ship or foreign flagship.

Mr. MENDELOWITZ. One of the observations we have with respect to shipping to Russia, is that the U.S.-flag ships have actually had very little experience shipping to Russia. They were not familiar with port conditions, customs, or operating in a country that has a number of things in a state of flux. Foreign-flag vessels that have regularly served the Russian ports have actually more experience in how to get through the—whatever obstacles they may face there. And so, in effect, the foreign-flag vessels, because of prior experience, may have had a somewhat easier time than the initial U.S.-flag ships.

Mr. ROSE. What about the first report on the Commission of Merchant Marine and Defense Funding finds—the conclusion that the Commission on Merchant Marine and Defense Funding, September 30, 1987, which—I guess you just basically think times have changed. Nobody said or gave the commission any evidence that would lead it to conclude that the United States should abandon its established policy of obtaining strategic sea lift resources from the merchant marines subsequently for a special quantity of limited shipping. You think times have changed?

Mr. MENDELOWITZ. Mr. Rose, we take as a given the objectives of the Merchant Marine Act of 1936. We take as a given the importance of the U.S.-flag fleet in meeting the needs of the Department of Defense, both with respect to manpower and ships as an emergency need arises.

What we are talking about in our report is not the validity of these objectives. What we are talking about in our report is the ability of cargo preference, when applied to the food aid programs, to contribute significantly to meeting those objectives.

And what we find is that a considerable amount of resources are being devoted to the application of cargo preference to food aid. And, unfortunately, this particular application of cargo preference has not made a significant contribution to meeting exactly the objectives that you are talking about.

Mr. ROSE. But you were focusing on food aid programs.

Mr. MENDELOWITZ. Correct.

Mr. ROSE. You were not taking a broader look at the value of the U.S. merchant marine and what it means to the defense of this country.

Mr. MENDELOWITZ. You are absolutely correct.

Mr. ROSE. Now, I think all members of agriculture and all farmers especially agree that they wish that somebody other than the U.S. Department of Agriculture's budget was the one that provided the resources for keeping cargo preference in this area. And maybe, just maybe, the Defense people are a little squeamish because they are worried that if they like it too much, you understand, it is too wonderful a program, that we will let them pay for it, you understand. So I think that kind of levels out the playing field here.

But I think that the cargo preference is a useful piece of a bigger Government policy that I totally support that says we should help maintain a strong U.S. merchant marine fleet. And I think your conclusions are very interesting and especially some of your recommendations.

Mr. Chairman, I thank you for giving me a little extra time.

Mr. PENNY. Mr. Allard.

Mr. ALLARD. Thank you, Mr. Chairman. I would like to request unanimous consent to insert my statement in the record.

Mr. PENNY. Without objection.

[The prepared statement of Mr. Allard follows:]

STATEMENT OF THE HONORABLE WAYNE ALLARD

SUBCOMMITTEE ON FOREIGN AGRICULTURE AND HUNGER

REVIEW OF THE GAO REPORT ON CARGO PREFERENCE REQUIREMENTS

SEPTEMBER 29, 1994

THANK YOU MR. CHAIRMAN. I AM PLEASED THAT YOU HAVE CALLED THIS HEARING AND THAT YOU REQUESTED THE GENERAL ACCOUNTING OFFICE TO LOOK AT CARGO PREFERENCE LAWS AND THE IMPACT ON AGRICULTURE PROGRAMS.

AS YOU KNOW CARGO PREFERENCE LAWS REQUIRE THAT 75% OF THE AGRICULTURAL COMMODITIES EXPORTED UNDER U.S. FOOD AID PROGRAMS, SUCH AS P.L. 480, BE CARRIED ON U.S. FLAG SHIPS. BECAUSE U.S. FLAG SHIPPING RATES ARE ALMOST ALWAYS HIGHER THAN FOREIGN FLAG RATES, CARGO PREFERENCE INCREASES THE COST OF PROVIDING THIS FOOD AID AND REDUCES THE AMOUNT OF COMMODITIES THAT CAN BE SHIPPED.

IN THE PAST USDA DESCRIBED CARGO PREFERENCE AS A MAJOR OBSTACLE IN ITS EFFORTS TO WORK OUT THE DETAILS OF LAST YEAR'S $700 MILLION RUSSIAN FOOD AID INITIATIVE. IN OUR SUBCOMMITTEE HEARING LAST YEAR, USDA EXPRESSED CONCERN ABOUT THE COST, AND IN PARTICULAR THE INCREASES IN COST, OF CARGO PREFERENCE.

ADDITIONALLY, USDA WAS CONCERNED THAT CARGO PREFERENCE PREVENTS FOOD AID SHIPMENTS FROM FOLLOWING NORMAL COMMERCIAL PRACTICES AND THEREFORE HINDERED USDA MARKET DEVELOPMENT EFFORTS.

I AM LOOKING FORWARD TO HEARING THE RESULTS OF THE GAO REVIEW OF CARGO PREFERENCE AND ITS RECOMMENDATIONS.

THANK YOU MR. CHAIRMAN.

Mr. ALLARD. Thank you. I have number of questions that I would like to ask.

I am rather interested in this phenomenon where most of our foreign aid shipments are clustered at the end of the year and then, because of that, they contribute to the additional cost of transportation. Can you give us—in the report here, you didn't seem to be able to go into much detail. But can you go into any detail as to why this occurs? Can USDA or AID do anything to change that occurrence?

Mr. MENDELOWITZ. We have some insights. However, quite honestly, we don't have a definitive answer to your question.

One reason why it gets clustered in the last half of the year is because it takes time to negotiate aid agreements with the recipient countries, and the commodities obviously can't be shipped until the aid agreements are reached and then implemented.

A second possible reason is, we are dealing with recipient countries that have limited storage capacity, limited port capacity, and if they fill warehouses with local-grown commodities during certain seasons of the year they may want the food aid cargo in a different part of the year. So that we don't think it is possible to fully even out shipments over the course of 12 months, but we do think—if effort and attention is devoted to this issue—to spread it out a little more evenly.

Second, if changes are made along the lines of the recommendations that we made, that would make it more likely that more efficient ships would be brought into the cargo preference trade if cargo preference is to be continued to be applied to food aid.

We have, we believe, another way to reduce costs even if cargos are bunched in a relatively short period of the year. One of the reasons why the costs get driven up is when you have peak demand inefficient ships get pulled in and find they can secure cargo. If we take away the 3-year rule and permit foreign-built U.S.-flag ships to immediately carry preference cargo, we could have a supply of more efficient ships which, in fact, would be available even during peak periods of demand to carry cargo at lower cost.

Mr. ALLARD. Do you have any idea what the additional transportation costs could be as a result of this clustering phenomena?

Mr. MENDELOWITZ. We don't have a quantitative estimate. I think, to answer that, I suppose we could go back and look at the rates charged during nonpeak periods and compare them with rates charged during peak periods. That would give us a ballpark estimate.

But even that is not a perfect estimate because if you have all of the current cargo bunched into one 6-month period the rates during that 6-month period are obviously going to be higher than they would be otherwise. But the rates during the first 6 months are going to be lower because there is a lot of competition for a relatively small amount of cargo.

So if you even out the flow more effectively throughout the year the rates during the slack period would be expected to go up some and the rates during the peak period would be expected to go down. But we don't have a good quantitative estimate for how much on a net basis might be saved.

Mr. ALLARD. The objectives of the Merchant Marine Act of 1936 are to maintain a U.S. merchant marine to serve as an auxiliary to the military, to carry a substantial part of the domestic and foreign commerce of the United States.

And, according to your report, the GAO report, U.S.-flag ships that are most dependent on Public Law 480 cargo are not necessarily viewed as militarily useful by the Department of Defense. And some of these ships apparently are the bulk carriers, tankers, and tug and barge combinations, and, apparently, they carry about 84 percent of the Public Law 480 cargo. Most of these ships apparently are not able to compete for foreign commercial cargos due to higher rates. How many ships are in this category and from what ports do they operate?

Mr. MENDELOWITZ. The ships that are primarily dependent on this particular cargo are roughly the 16 to 18 bulk carriers that basically operate full time in the food aid cargo preference trade. In addition, there are some tug/barge combinations which aren't militarily useful because they just can't make the speed required by DOD.

The other types of ships—for example, you mentioned tankers—that do take some of the cargo preference food aid generally tend to be ships that take a very small amount of it, and generally the cargo preference trade represents a very small amount of those ships' revenues. They may have a slack month, or they will jump into the cargo preference trade during the peak time when demand and rates are very high, and then go back to carrying petroleum.

So the ships that really are dependent on this trade are, as I indicated, these 16 to 18 bulk carriers, and they operate primarily out of the gulf and the east coast.

Mr. ALLARD. OK, the question, do you have any idea on how many—you have 16 or 20 bulk carriers and then the tankers and tug/barge combinations. Do we have any idea on those?

Ms. HOFFMAN. We found that over the last 3 years, there were about 25 tankers that had carried at least one food aid cargo during that time. Very few of those 25 carried any significant amount of food aid. There also were about 71 or so tug/barge combinations that had carried at least one cargo during that time frame. It has varied as to how much those particular types of vessels are dependent on these cargos. Some carry a lot, some carry very little.

Mr. MENDELOWITZ. Some of the tug/barge combinations, for example, are used to provide liner service in the Caribbean. So this represented a small part of their general trade.

Mr. ALLARD. I see. You are recommending in the GAO report that the Secretary of Transportation prove the efficiency of ships that carry U.S. food aid. What way do you think we can improve the efficiency of these ships?

Mr. MENDELOWITZ. There are certain costs that are controllable and certain costs that aren't. U.S. seamen receive wages substantially above wages of. Third World country sailors. The wage bill of U.S. ships is higher because we have a country in which our wage rates are higher and the standard of living of our people is higher.

I don't think anyone is suggesting that we ask U.S. seamen to work for wages that are appropriate in a Third World country. But, with that said, if you look at the age and operating characteristics of the ships dependent upon food aid preference cargo, they tend to be quite old and inefficient when compared to newer ships.

Newer ships require fewer berths, fewer seamen to operate them. They operate much more efficiently in terms of the amount of fuel used to move them. They can be loaded more quickly and off-loaded more quickly so you have less down time in ports and you can, in fact, spend more of a year sailing the high seas carrying cargo rather than loading and off-loading in ports.

So that if, in fact, the trade were to include ships that were newer and more efficient in terms of design, costs could be reduced because crews would be a little smaller. The ships would be more fuel efficient. They would be able to load and off-load more quickly, more efficiently than the current ships.

Mr. ALLARD. What is this cost of providing subsidies to the flag-ships—total U.S. subsidies to U.S.-flag ships?

Mr. MENDELOWITZ. Well, under the application of cargo preference for food aid programs, in round numbers, about $200 million a year was spent over the preceding 3 years. So, over a 3-year period, it was in round numbers something on the order of $600 million.

Mr. ALLARD. But there's other subsidies, too. There is construction subsidies, aren't there? And operating subsidies?

Mr. MENDELOWITZ. The Congress, through legislation over the years, has created a number of programs to help sustain the U.S.-flag fleet to ensure that national interest and defense needs were met.

The United States did operate a construction subsidy program to bring down the cost of U.S.-built ships to purchasers, but this program ended in 1982; and, since 1982, there have been no construction subsidies for ship construction.

I understand there is a new proposal that would seek to revitalize the U.S. shipbuilding industry, and there might be some subsidies associated with that. However, the United States is a party to a new international agreement which seeks to phase out construction subsidies for ships in the coming years.

Mr. ALLARD. So do you have any figures when you take all these subsidies and total as to what they would amount to? I think that would be helpful. If you don't, let the record note that you do not know, but you will provide those for us.

Mr. MENDELOWITZ. I am sorry that we don't have the number available because we really focused on the application of cargo preference food aid. We focused on the costs associated with that piece of the program. We have not tried to sum up the cost of all cargo preference or the cost of all subsidies to the maritime industry. That was well beyond the scope of this particular study.

Mr. ALLARD. Well, if you could get those, it would be helpful. And I think to the committee, if you possibly can.

[The information appears at the conclusion of the hearing.]

Mr. ALLARD. Let me see, food aid transportation costs are increased about 75 percent because all shipments must be on U.S.-flag carriers. For the period from 1991 through 1993, GAO has attributed about $600 million as additional cost for transportation.

How much additional U.S. commodities could have been provided to the needy countries had the expenditures not been required?

Mr. MENDELOWITZ. Mr. Allard, I have to be quite honest. There is, as I understand it, a debate in Congress as to exactly which beneficiary is, in effect, represented by which particular appropriation. While the appropriation for the food aid program includes an appropriation to cover the cost of cargo preference, it is unclear to us what the appropriation for the food aid program would be without cargo preference, whether it would be the same amount or, in fact, would be less. And without having that guidance from the Congress, it would be hard to determine how much additional food aid could have been shipped.

Mr. ALLARD. Do any recipient countries select commodities based on the cost of transportation by buying cheaper commodities because of the higher transportation cost to us?

Mr. MENDELOWITZ. We found that the application of cargo preference has, in fact, altered the types of commodities that certain countries purchase as well as the cost of the commodities that they purchase.

If you are a purchaser in a foreign country and you have a fixed amount to spend and you are interested in maximizing the amount of food that you receive with that aid appropriation or you are interested in buying a particular commodity with that aid appropriation, you will look around to try to, I think, get what you want at the lowest cost per ton, where the cost per ton includes both cost of the commodity as well as the cost of shipping.

Unfortunately, some countries were unable to assess whether in fact they were getting the lowest cost per ton because decisions in the end were dictated by the availability of U.S.-flag ships, not by the particular commodity that they wanted to buy. We have a number of examples where, in fact, the application of cargo preference in a sense forced countries to purchase commodities at a higher cost per ton or to purchase commodities that were not the preferred first-choice commodity.

For example, Western White wheat is a commodity that a certain purchaser wanted to purchase. They couldn't get it because there was no U.S.-flag ships capable of carrying it from the western ports. And because there were no U.S.-flag ships capable of carrying from the western ports, they weren't able to make a comparison what the landed cost would be for that particular commodity relative to an alternative.

The U.S. Department of Agriculture told us it is these types of restrictions that keep Public Law 480 from being used as an effective marketing tool because they can't grow markets in commodities that some potential customers may want because of the inability of shipping from ports where those commodities are available.

Mr. PENNY. If I might interject, is there any empirical evidence that because of the cargo preference costs recipient countries negotiate a lower commodity price as a way of reducing the overall cost of the shipment?

It is slightly different than Mr. Allard's question which is, do they purchase a less costly commodity that maybe serves their same food need? I am asking whether they buy the same commodity but drive a harder bargain and lower the price of that commodity because they also have to accommodate these transportation costs.

Mr. MENDELOWITZ. The volumes involved in cargo preference are fairly small, and, because of that, the recipients in effect are price takers. The amounts involved are too small to really give them the leverage to affect the price in a material way. But with that said——

Mr. PENNY. So they are price takers on the grain and they are also left with little recourse on the shipping costs. They are forced to do that, as well.

Mr. MENDELOWITZ. That is correct, given that the shipping costs put out for bid. If it is a period of peak demand, there is going to be limited ships available. And they—in effect, as long as a U.S.-flag vessel is willing to carry the cargo at guideline rate or less, they are price takers.

What I was going to say, though, in response to your question, is that a rational foreign recipient should want to receive commodities at the lowest landed cost at the port of destination which, when you consider includes both the cost of the commodity and cost of shipping, in effect there is a trade-off.

So if you are a price taker, you can buy a lower-cost commodity in one place or higher-cost commodity in another place. The transportation cost could affect the decision as to which one you want because a higher-cost commodity could have a lower landed cost if the transportation cost from the port of embarkation is less.

Mr. ALLARD. There is another interesting phenomena that I want to bring up for discussion in this subcommittee hearing, and that is apparently the USDA and the AID has imposed additional requirements on shipowners where they are in charge of inland storage and fumigation. Would you go into some detail on that? Because I could see that would be a real problem for shipowners when they are required to take it inland, not even part of their business, and I wonder how they go about these calculations.

Mr. MENDELOWITZ. I think you have identified one of the factors that adds to the cost of the program. Commercial terms as a general rule don't include things like fumigation of the cargo or inland transportation. Those particular aspects of the delivery of the commodity are typically the responsibility of either the seller or the purchaser. To the extent to which——

Mr. ALLARD. Would you repeat that? You say you saddled that on the purchaser?

Mr. MENDELOWITZ. It would typically be the responsibility of the purchaser or the seller.

Mr. ALLARD. Right.

Mr. MENDELOWITZ. And because of the way these programs are designed, the U.S. aid agencies are interested in providing additional assistance to the beneficiary countries. They have required that the shippers provide these services.

Mr. ALLARD. Have they done that on their own without any authorization from Congress or they just feel like this is necessary? Or how has that come about?

Mr. MENDELOWITZ. They have the authority to do that.

Mr. ALLARD. They have the authority.

Mr. MENDELOWITZ. Yes. So what happens is these types of costs, such as inland transportation, fumigation costs, are costs that, in effect, are not things that the shipowner can control. The shipowner must contract with someone else to provide those services.

And, in fact, when the shipowner bids on a cargo, the shipowner includes both the guideline rate—or I should say MARAD or other U.S. agencies include both the guideline rate as well as the imputed cost of these additional services in determining the maximum cargo rate that may in fact be charged.

And because these costs represent a noncontrollable in effect for the shipowner, the shipowner, in order not to lose money, must be sure to estimate these costs correctly. And the way to avoid losses, I would think, is to estimate very conservatively, that is on the high side, so that the shipowner won't get stuck with a bill that far exceeds what he/she is being compensated for.

So the incentives in this system are incentives that work toward raising the cost of providing the services rather than minimizing the costs.

Mr. ALLARD. Have you thought about any other alternatives in your report? How you might get that grain imported other than through that mechanism? Or do you know that is the most efficient way to have it, get it done?

Ms. HOFFMAN. What we have really suggested in our report is that the food aid agencies look at experimenting with different means to do that. Their alternatives could be reimbursing the recipient country for that particular item.

But we are not imposing or suggesting anything specific. We are asking them to really look at it and experiment with it because each recipient country has a unique situation, and the services you are trying to provide to the food aid program have different objectives.

But because of the cost savings potential to doing things differently instead of always providing services through the shipping rates and through the shipowners, we are asking them to consider looking at those terms that are more consistent with those used for commercial cargos.

Mr. ALLARD. Do we have legislation that allows the Department of Agriculture to try out these various alternatives?

Ms. HOFFMAN. My understanding is that they are allowed to pay for these services in any way. I mean, their funds can be used to pay for them. The terms by which they are paid for is up to them.

Mr. ALLARD. So there is discretion given to the Department of Agriculture to try these other approaches.

Ms. HOFFMAN. Yes.

Mr. ALLARD. OK. Thank you, Mr. Chairman. That completes my questioning.

Mr. PENNY. Thank you, Mr. Allard.

I want to return to this question of shipments to Russia. It is my understanding that during sort of the peak demand period where roughly 73 percent of the shipments did not go on U.S.-flag vessels. Was that just an availability issue or were there other factors that allowed us to circumvent the cargo preference requirement?

Mr. MENDELOWITZ. Well, the shipments to Russia that you refer to took place during the period of peak demand for U.S.-flag ships under the cargo preference program, and there basically was just very little availability of U.S.-flag ships. And there is a provision in the cargo preference law which basically says cargo preference is to be applied if U.S. ships are available at fair rates. And, basically, U.S. ships were just not available.

Mr. PENNY. So as long as U.S. ships are available, there is no getting around it. The possibility has been posed to me that we may have recipient countries that would find it much easier to take shipments from western ports except that there are no U.S. vessels suitable for those shipments presently at our western ports. Are these countries forced to take shipments from eastern ports?

Mr. MENDELOWITZ. Generally, yes.

Mr. PENNY. They are.

Mr. MENDELOWITZ. The problem with the western ports is the classic chicken and egg problem. If there were sufficient food aid cargos from the western ports, U.S.-flag ships would be there. But the amount of food aid cargos that could be shipped from the western ports are fairly small.

Mr. PENNY. So proximity to the western ports is not a condition that would allow the cargo preference law to be circumvented. A country couldn't say we are closer to those ports. If there are no ships there, we are going to move it on foreign-flag vessels.

The U.S. Government basically says, well, then we may not have ships there but we have them available in the eastern ports and you are going to wait on the U.S.-flag vessels for this shipment?

Mr. MENDELOWITZ. In terms of how the program has been applied up to now, that is correct.

In addition, I will give you another example consistent with that. That is, the lack of availability of exactly the type of ship you want is also not a reason for circumventing cargo preference requirements.

Mr. PENNY. Which then requires what? A wait until the ship is available or taking the ship—a ship that isn't suitable and making it work somehow?

Mr. MENDELOWITZ. It means using a ship that can carry your cargo but is not most suitable.

Mr. PENNY. Right, like oil?

Mr. MENDELOWITZ. Maybe. An example is tractors or large earthmoving equipment, which is U.S. Government impelled cargo and covered by cargo preference. This particular type of item may be shipped most efficiently by "RoRo" ship. With such a ship you drive the tractor on, get to your destination, and drive it off. However, a U.S.-flag "RoRo" ship may not be available.

Mr. PENNY. So you use a crane ship instead.

Mr. MENDELOWITZ. You use a crane. And, you may have to partially disassemble the tractors, store the parts separately, and then reassemble it all at its destination.

Mr. PENNY. How are those costs handled if there is disassembly and reassembly and all the rest? Is that just another add-on to the recipient?

Mr. MENDELOWITZ. I will be honest. I just don't remember whether they are borne by the exporter or covered by cargo preference. I just don't know. I would have to check on that.

Mr. PENNY. Let me move to one last question and then we will wrap things up for the morning.

Your estimate on the annual cost of cargo preference is $200 million and that applies to roughly 75 percent of our Public Law 480 shipments. Title III of Public Law 480, as you know, deals with grants to the poorest countries. It is an alarming statistic to me to see that the total appropriation set aside for fiscal year 1995 for these poorest countries under title III is only $157.4 million. I am just looking for an average number here. How much of that $157.4 million would likely be consumed in shipping costs under the cargo preference requirement?

Ms. HOFFMAN. Well, what we found for the period——

Mr. PENNY. And, again, I know this varies from country to country and all the rest, but roughly how much of that $157 million will we lose in commodities simply because of the shipping costs? I am looking to the higher shipping costs that are attendant to the cargo shipping law.

Ms. HOFFMAN. For the period we looked at the additional cost, the OFD portion consumed about 9 percent of the total program. And then the transportation costs were—of U.S. flag transportation costs were equivalent to 13 percent of your entire program. That is for all of the Public Law 480 programs together. I can't tell you specifically for title III at this point.

Mr. MENDELOWITZ. We can try to get that for the record, if you would like.

Mr. PENNY. If we can break that down, I would like to take that $157.4 million figure and give me the best estimate you can based on prior experience as to how much of that will go toward commodities, how much of that will go toward shipping and then how much of that shipping cost is really the higher cost that we must pay due to the cargo preference requirement.

Because, clearly, of all the Public Law 480 categories, this is the most critical, and it is heartbreaking to me to see a large volume of potentially lifesaving commodities denied because we are using the money instead to pay higher-than-necessary shipping costs on these shipments.

Mr. MENDELOWITZ. Mr. Penny, we have an example in our report on page 50 that discusses a title III grant to Sri Lanka in——

Mr. PENNY. That is one and those costs——

Mr. MENDELOWITZ. Yes. That, I think, gives you a ballpark estimate of the proportions that you are looking for.

Mr. PENNY. Yes. There are 52 percent of the transportation costs that result as additional cost as the result of cargo preference.

Mr. MENDELOWITZ. Right. We have a situation in which Sri Lanka received just under $59 million to purchase wheat and pay for its transportation for the cargo preference year ending in March of 1994. And with these funds Sri Lanka purchased seven lots of wheat. And roughly $20.6 million or 35 percent of the $59 million in allocation was spent on transportation and not available to purchase wheat. And out of the $20.6 million, $10.8 million or about 50 percent of the transportation costs were spent on OFD. So, in this case, one-third of the allocation went for transportation and then one-half of that amount represented OFD. So, in effect, you are talking about 16, 17 percent of the allocation going to cover the added cost of cargo preference. If you extrapolate that to your—what was it? $150 million?

Mr. PENNY. $157 million.

Mr. MENDELOWITZ. $157 million, you are talking about something roughly on the order of $25 million would be spent on OFD if you can extrapolate from the Sri Lanka example to the total budget.

Mr. PENNY. Thank you. I appreciate that response.

And, again, we thank you for your appearance this morning. Your testimony was quite helpful. The report, as usual, is outstanding.

And I want to say, before we wrap up this morning's hearing, a fond farewell to GAO. I have enjoyed our working relationship over the years and have the utmost respect for your work and the staff that labors so hard to put together these reports. As Mr. Barlow said earlier, we wouldn't know what to do without the GAO's assistance.

I yield to Mr. Allard at this point.

Mr. ALLARD. Mr. Chairman, I don't know whether this is going to be our last hearing.

Mr. PENNY. Wait a minute. Let me say thank you to Mr. Allard and to the committee members for their interest in these issues over the past 2 years and to your staff and to the minority staff as well as my staff and the majority staff. It has been great working with all of you.

Mr. ALLARD. Well, I was sort of preempted here a little bit. I wanted to thank you for the great job you have done as chairman of this subcommittee. It has been a pleasure to work with you and your staff, and I think that the meetings that you have conducted here have been very fair to both sides. I think today is a good example of fairness that we have tried to bring out in our discussion. And so I just want to wish you the very best in whatever future endeavors, and we will miss working with you in the coming years. Thank you.

Mr. PENNY. With that, the meeting stands adjourned.

[Whereupon, at 11:10 a.m., the subcommittee was adjourned, to reconvene subject to the call of the Chair.]

[Material submitted for inclusion in the record follows:]

CARGO PREFERENCE REQUIREMENTS:
OBJECTIVES NOT MET WHEN APPLIED TO FOOD AID PROGRAMS

SUMMARY OF STATEMENT BY ALLAN I. MENDELOWITZ, MANAGING DIRECTOR
INTERNATIONAL TRADE, FINANCE, AND COMPETITIVENESS
GENERAL GOVERNMENT DIVISION

"Cargo preference" is the reservation of government-financed or -sponsored oceangoing cargos specifically for U.S.-flag ships. Due to cargo preference requirements, the government has incurred almost $600 million in additional transportation costs associated with transporting food aid tonnage on U.S.-flag ships, rather than on lower-cost foreign flag ships, over the past 3 years.

The objective of cargo preference is to help maintain a U.S. merchant marine (1) to serve as a naval and military auxiliary in times of war or national emergency and (2) to carry a substantial portion of U.S. domestic and foreign waterborne commerce. However, applying cargo preference requirements to U.S. food aid programs contributes little to meeting this objective. The bulk commodity-carrying U.S.-flag ships that transport the majority (84 percent) of food aid tonnage--largely as a result of cargo preference requirements--are not viewed as militarily useful by the Department of Defense (DOD). Furthermore, as for the crews that support those ships, DOD believes that they could be a potential source of manpower for the Ready Reserve Force but does not believe that applying cargo preference to food aid programs is a cost effective means of providing for crews. The U.S.-flag ships that DOD does view as militarily useful, and which carry some food aid cargos (16 percent), are liners.

In addition, all domestic waterborne commerce, i.e., cargos shipped between U.S. ports, are required to be carried by U.S.-flag ships under the Merchant Marine Act of 1920, commonly referred to as the "Jones Act." As for foreign commerce, U.S.-flag ships now carry only about 4 percent of all cargos imported into or exported out of the United States, and food aid cargos account for less than one-fourth of that 4 percent.

GAO also determined that cargo preference requirements adversely affect operations of the food aid programs. Given the findings, GAO suggests in its report that Congress consider a more efficient alternative for helping support the U.S. merchant marine. GAO also recommends ways to reduce food aid transportation costs should Congress choose to continue applying cargo preference requirements to food aid programs.

Mr. Chairman and Members of the Subcommittee:

I am pleased to be here today to testify before this Subcommittee on the results of our review of the application of cargo preference laws to U.S. food aid programs. During each of the past 3 years, an average of about $200 million in government funds have been used to pay the added cost of shipping U.S. food aid to foreign countries on U.S.-flag ships rather than on lower cost foreign-flag ships. This cost has been incurred to meet the cargo preference requirement that at least 75 percent of U.S. food aid tonnage be transported on U.S.-flag ships. Although it results in higher transportation costs for U.S. food aid programs, the objective of this requirement is to help ensure that an adequate and viable merchant marine is maintained (1) to serve as a naval and military auxiliary in times of war or national emergency and (2) to carry a substantial portion of U.S. domestic and foreign waterborne commerce.

My testimony today will provide an overview of our findings--which are more fully discussed in our report on this subject that we issued today to this Subcommittee and to the Subcommittee on Merchant Marine, Senate Committee on Commerce, Science, and Transportation.[1] I will discuss the (1) extent to which applying cargo preference requirements to U.S. food aid programs

---

[1]See CARGO PREFERENCE REQUIREMENTS: Objectives Not Significantly Advanced When Used in U.S. Food Aid Programs (GAO/GGD-94-215, September 29, 1994).

is meeting the intended objective, (2) how cargo preference requirements affect U.S. food aid programs, and (3) how the practices currently used by the U.S. Department of Agriculture (USDA) and the Agency for International Development (AID) to manage food aid transportation affect its cost.

BACKGROUND

Currently available data show that in 1991, U.S.-flag ships transported about 4 percent of all waterborne commerce imported into or exported from the United States (foreign commerce). Food aid preference cargos accounted for less than one-fourth of that 4 percent.

Food aid tonnage also accounts for a small portion of all U.S. agricultural exports. For example, during fiscal years 1991-93, U.S. food aid tonnage represented 6.7 percent of all U.S. agricultural export tonnage and 18.1 percent of all U.S. agricultural tonnage exported with assistance from the U.S. government. The value of all agricultural commodities exported from the United States for fiscal years 1991 through 1993 totaled \$122.4 billion, with \$98.1 billion, or 80.2 percent, in private sector commercial sales; \$19.2 billion, or 15.7 percent, in government-sponsored credit sales in which financing was provided at "near commercial" rates and terms; and \$5.1 billion, or 4.1 percent, in U.S. food aid programs.

Since the passage of the Cargo Preference Act of 1954,[2] Congress has required that a certain percentage of U.S. food aid be transported on U.S.-flag ships. This requirement is known as "cargo preference." Initially, the requirement was that 50 percent of all food aid tonnage be transported on U.S.-flag ships. Beginning in 1986, this requirement was gradually increased to 75 percent. The food aid programs subject to this requirement include those carried out under the Agricultural Trade Development and Assistance Act of 1954,[3] as amended, widely know as "Public Law (P.L) 480," and under section 416(b) of the Agricultural Act of 1949,[4] as amended. Cargo preference also applies to assistance provided under the Food for Progress program, which was enacted as part of the Food Security Act of 1985.[5]

In addition to cargo preference, the Merchant Marine Act of 1936,[6] as amended, established subsidy programs to help support the U.S. merchant marine. Congress developed these programs to

---

[2]The Cargo Preference Act of 1954 (Ch. 936, 68 Stat. 832, Aug. 26, 1954), as amended by the Food Security Act of 1985 (Public Law 99-198, Dec. 23, 1985), amended section 901(b) of the Merchant Marine Act of 1936 (Ch. 858, 49 Stat. 1985, June 29, 1936) to require that at least 75 percent of U.S. food aid tonnage be shipped on U.S.-flag commercial ships.

[3]Ch. 469, 68 Stat. 454, July 10, 1954.

[4]Ch. 792, 63 Stat. 1051, 1058, October 31, 1949.

[5]Public Law 99-198, December 23, 1985.

[6]Ch. 858, 49 Stat. 1985, June 29, 1936.

respond to general downturns in the U.S. maritime industry and to help support a U.S. merchant marine sufficient to meet the objectives of the Merchant Marine Act of 1936. One of these subsidy programs helped offset the high costs of constructing ships in U.S. shipyards, although no funding has been provided since 1982. Any substantial new funding under this subsidy program is unlikely in the near future as the United States has entered into an agreement with the Organization for Economic Cooperation and Development to eliminate government subsidies for shipbuilding by January 1, 1996. Another subsidy program helps U.S. shipowners compete with their foreign competitors by offsetting high U.S. operating costs. Current subsidies under this program will expire by December 1998. However, the administration has submitted proposed legislation to Congress that would continue such support by establishing a similar subsidy program.

Yet even with the support provided by these programs, the status of the U.S. merchant marine has continued to decline. The Maritime Administration (MARAD) reported that between 1982 and 1992, the number of U.S.-flag privately owned ships decreased from 574 to 394 (a decrease of 31.4 percent). During this same period, the deadweight tonnage capacity[7] of these ships

[7]Deadweight tonnage is the total carrying capacity of a ship expressed in tons of 2,240 pounds. Carrying capacity is the difference between the displacement of the empty ship and the displacement of the ship fully loaded.

decreased from 21.5 million tons to 19.7 million tons (a decrease of 8.5 percent). In addition, the average monthly U.S. seafarer employment decreased 11.3 percent--from 16,308 in fiscal year 1991 to 14,466 in fiscal year 1992.

From cargo preference years[8] 1991 through 1993, 144 privately owned U.S.-flag ships carried food aid preference cargos. These 144 ships included 99 liners, 25 tankers, and 20 bulk carriers. In addition, 71 tug/barge combinations, which MARAD does not include in its count of U.S.-flag ships, also carried food aid preference cargos during this time. These privately owned U.S.-flag ships and tug/barge combinations carried about 17.1 million tons of food aid from cargo preference years 1991 through 1993. A majority (84 percent) of this food aid tonnage was bulk commodities such as wheat or corn and was shipped on bulk carriers, tug/barge combinations, and tankers that were typically chartered for specific voyages. The remaining food aid tonnage (16 percent) was processed products such as cans of vegetable oil or bags of flour or rice. This tonnage was transported on ships that provided a regularly scheduled service between specific

---

[8]A cargo preference year spans from April 1 of one year to March 31 of the next year and was created to measure compliance of food aid cargos with the cargo preference requirements spelled out in the Food Security Act of 1985. The act required that the amount of food aid tonnage transported on U.S.-flag ships be gradually increased from 50 percent of all food aid tonnage shipped in 1985 to 75 percent by April 1988. The act also required MARAD to fund the additional cost of using U.S.-flag ships--the ocean freight differential (OFD)--for the additional 25 percent of food aid shipped on U.S.-flag ships.

ports known as "liner" service.

OBJECTIVES OF MERCHANT MARINE
ACT OF 1936 NOT SIGNIFICANTLY
ADVANCED BY FOOD AID PREFERENCE CARGOS

The application of cargo preference to food aid programs does not significantly contribute to meeting the intended objectives of the Merchant Marine Act of 1936, as amended.

The Contribution of Food Aid Preference Cargos
to Maintaining a U.S. Merchant Marine as a
Military and Naval Auxiliary

While applying cargo preference requirements to food aid programs does help support some U.S.-flag ships and their crews, these ships are not generally viewed by the Department of Defense (DOD) as being militarily useful in terms of serving as a naval and military auxiliary. DOD officials indicated to us that the U.S.-flag ships that transported 84 percent of food aid cargo preference tonnage--bulk carriers, tankers, and tug/barge combinations--would be used for military or military support purposes only as a last resort. While these U.S.-flag ships may have been military useful at one time, DOD officials explained that this view has changed because of DOD's changing national security needs, the shift in the shipping industry to the use of containers, and the increase in the size and weight of military equipment.

Those U.S.-flag ships that DOD said are militarily useful and that participate in the food aid programs are those engaged in providing liner service. DOD officials told us that these ships are a more efficient way to transport ammunition, equipment, and supplies. DOD officials also told us that liners will become more useful as DOD continues to expand its use of containers for packaging supplies and equipment for ocean transport.

In addition, while the crews on bulk carriers, tankers, and tug/barge combinations that transport food aid could be used to help crew the Ready Reserve Force,[9] DOD does not believe that applying cargo preference to food aid programs is a cost effective means for providing for crews. However, DOD officials also recognize that as the number of U.S.-flag ships--and their crews--continues to decrease, and the number of ships in the Ready Reserve Force increases, they may need to consider options such as a merchant marine reserve program to ensure that adequate crews are available in the future.

MARAD officials disagreed with DOD's assessment of its need for the U.S.-flag ships and crews supported by food aid preference cargos. MARAD said that food aid preference cargos are very

---

[9]The Ready Reserve Force is a specific component of the National Defense Reserve Fleet. Ships in the Ready Reserve Force are to be kept in a state of readiness that enables them to be activated in 4, 5, 10, or 20 days to meet military sealift surge requirements in event of war or emergencies. MARAD maintains these ships for DOD use.

important to the support of a significant number of U.S.-flag ships and crews. MARAD said it views these ships as militarily useful because they were built to DOD specifications in order to fulfill some military purpose. While DOD officials agreed that the ships in question could be used for military purposes, they also said that given their current needs, these ships would be one of their last choices to transport military equipment. The investment required to modify them to carry military supplies and equipment would be too high, they said, and the ships could only be loaded or unloaded in modern ports, given the type of gear that would be required to handle heavy military equipment. In addition, while DOD recognized its future needs for crew for the Ready Reserve Force, it said that it is not an efficient use of resources to provide for these needs through supporting U.S.-flag ships with food aid preference cargos.

The Contribution of Food Aid Preference Cargos to Maintaining a U.S. Merchant Marine to Carry a Substantial Portion of U.S. Waterborne Commerce

The Merchant Marine Act of 1920,[10] commonly referred to as the "Jones Act," restricts domestic waterborne commerce to U.S.-constructed, U.S.-flag ships and requires that U.S.-flag ships carry all domestic waterborne commerce of the United States. Therefore, food aid cargo preference requirements are not

[10] Ch. 250, 41 Stat. 988, June 5, 1920.

relevant to satisfying this objective of the Merchant Marine Act of 1936.

Food aid cargo preference requirements are relevant in the case of foreign commerce, as they reserve 75 percent of all U.S. food aid tonnage exclusively for U.S.-flag ships. In addition, operating differential subsidies help U.S.-flag ships carry foreign commerce by placing their operating costs on a parity with those of their foreign competitors. However, U.S.-flag ships carry only about 4 percent of all U.S. export and import foreign commerce, and less than one-fourth of that is food aid preference cargos. Therefore, food aid preference cargos contribute little to the objective of maintaining U.S.-flag ships to carry a substantial portion of U.S. foreign commercial cargos.

Dependence of U.S.-Flag Ships on Food Aid Preference Cargos Varies

Most U.S.-flag bulk carriers and tug/barge combinations, which typically do not receive or use operating differential subsidies, are virtually dependent on food aid preference cargos to operate as U.S.-flag ships because they are unable to compete successfully for other commercial cargos in foreign commerce. Almost all U.S.-flag tankers that carry bulk food aid preference cargos do receive annual operating subsidies to help them compete for foreign commercial cargos. However, since these subsidies are expiring, tankers have, and probably will, become more dependent on food aid preference cargos. U.S.-flag liners also carry some food aid tonnage but are able to compete successfully for foreign commercial cargos because they either operate with operating differential subsidies,[11] operate as part of an ocean liner conference,[12] or do both. While U.S.-flag liners carried 16 percent of the food aid tonnage transported on U.S.-flag ships from cargo preference years 1991 through 1993, for a majority of these liners--about 75 percent--food aid preference cargos made up no more than 4 percent of their cargos, and were not the reason why they maintained their U.S. flag status. For the other 25 percent of the liners, food aid preference cargos made up between 5 and 40 percent of their cargos. In addition to food aid cargos, U.S.-flagged ships carried other U.S. government preference cargos, or domestic or foreign commercial cargos.

U.S. Laws and Regulations Affect Shipping Costs

Without operating differential subsidies, U.S.-flag ships generally cannot successfully compete for foreign commercial

---

[11]U.S.-flag liners that receive operating differential subsidies may carry food aid preference cargos and still receive their subsidy because food aid cargos typically represent such a small portion of their total cargo.

[12]Liner companies that serve the United States may be members of cartels, called "ocean freight-rate conferences." Members of such conferences often have agreements on (1) the freight rates they charge (as a way to restrict competition), (2) their sailing schedules and ports of call, and (3) the pooling of cargos or revenues. Conferences that serve the United States must file tariffs with the Federal Maritime Commission that state their rates, terms, and conditions of transport covering all commodities they propose to carry for the general public.

cargos. According to MARAD and the U.S. shipowners we talked to, this is due, in part, to the costs associated with complying with the U.S. laws and regulations governing U.S.-flag ships. These costs help explain why U.S.-flag shipping rates for the food aid programs are as much as twice foreign-flag shipping rates for comparable ships.

While the benefits that accrue from these U.S. laws and regulations--U.S. citizen employment, protection of the environment, and support of U.S. shipyards--are significant, according to the U.S. shipowners we talked to who have carried food aid preference cargos, their cost impedes their ability to compete with foreign-flag ships. The most costly U.S. laws and regulations are those related to ship construction, ship maintenance and repair, and crewing requirements.

U.S.-flag ships are required to be constructed to U.S. Coast Guard safety standards, whether they are built in a U.S. or a foreign shipyard. These standards are more stringent than international standards, which adds to the construction cost of U.S.-flag ships. U.S. law also requires that owners of U.S.-flag ships either maintain or repair them in a U.S. shipyard, whose services are more expensive than those available at a foreign shipyard, or pay a 50-percent U.S. Customs duty on the value of work done in a foreign shipyard. Even with the added cost of the 50-percent duty, U.S. shipowners told us that they still find it advantageous, in most circumstances, to have maintenance work on their ships done in foreign shipyards.

In addition, U.S.-flag ships are required to employ U.S. citizen crews. This requirement greatly increases shipowners' operating costs because wages and benefits paid to U.S. crews generally far exceed those provided to crews of foreign-flag ships. For example, according to a November 1993 article in the Journal of Commerce, monthly crew costs for a U.S.-flag liner can top $310,000, with the captain receiving $44,000 a month in wages and benefits. A similar foreign-flag ship spends about $100,000 a month for its crew, with the captain receiving about $10,000. The article did not specify the countries in which the foreign-flag ships used for this comparison were registered. In addition, the shipping companies we interviewed who have foreign-flag ships similar to the U.S.-flag ships they use to carry food aid preference cargos said that the daily crew costs for their U.S.-flag ships are at least three times those of their foreign-flag ships.

Moreover, U.S. shipowners whose ships carry food aid preference cargos do not have many incentives to lower their costs. For example, one of the primary ways that U.S. shipowners can reduce their operating costs is to invest in newer, more efficient ships. U.S. shipowners are discouraged from doing this because of the high cost of constructing ships in U.S. shipyards. Also, current legislation requires that ships constructed in less expensive foreign shipyards be operated as U.S.-flag ships for 3 years before they are eligible to carry food aid preference cargos. If a U.S. shipowner were to purchase a foreign-built or rebuilt bulk carrier, it would be almost impossible to operate that ship as a U.S.-flag ship for 3 years without food aid preference cargos because the costs would be too high to be price competitive compared to foreign-flag competition. Furthermore, since there is a limited number of U.S.-flag ships available to carry these cargos, and MARAD's "fair and reasonable" guideline rates[13] establish rate ceilings for each individual ship based on its actual costs irrespective of its efficiency, shipowners are able to secure food aid preference cargos despite the high cost of operating their older, inefficient U.S.-flag ships.

CARGO PREFERENCE LAWS ADVERSELY AFFECT U.S. FOOD AID PROGRAMS

Cargo preference laws can adversely affect the operation of U.S. food aid programs. The most significant effect of applying cargo preference to food aid programs is the additional costs associated with using U.S.-flag ships to transport food aid. From fiscal years 1991 through 1993, almost $600 million in government funds have been paid to U.S. shipowners in ocean

[13]MARAD is responsible for calculating guideline rates for food aid preference cargos to ensure that U.S.-flag shipping rates are "fair and reasonable."

freight differential. Ocean freight differential is the difference between the cost of the less expensive foreign-flag ships and the U.S.-flag ships used to carry food aid cargos. Generally, USDA pays two-thirds of this cost, and MARAD pays one-third. This $600 million in ocean freight differential represented 9 percent of all food aid program funds and 34 percent of all food aid program funds spent on transportation for fiscal years 1991 through 1993.

For several of the food aid programs, the amount spent on ocean freight differential reduces the amount available to purchase commodities. For example, for the P.L. 480 title III program, the funds available for each country must be used to purchase both the commodity and its transportation. While the estimated cost of transporting food aid is included in the total amount provided to each recipient country, in many instances the amount spent to transport food aid on U.S.-flag ships can exceed this estimate. Therefore, the amount spent on U.S.-flag transportation can directly affect the amount of commodity that can be purchased.

This situation occurred for Sri Lanka's wheat purchase made during the cargo preference year ending March 31, 1994. Sri Lanka received $58.9 million to purchase wheat and its transportation. Due to the high rates charged by U.S.-flag ships, 35 percent, or $20.6 million of the funds, was spent on transportation and therefore was not available to purchase wheat. About $10.8 million, or 52 percent of these transportation costs, was spent on ocean freight differential. AID officials explained that Sri Lanka's experience was significant because the amount spent on the ocean freight differential was more than expected and reduced the amount of wheat Sri Lanka had hoped to purchase. AID officials said they believed the reason that these differential expenditures were so high for Sri Lanka was because its shipments were competing for the U.S.-flag ships capable of carrying 50,000 metric tons with those shipments sent to Russia under the Food for Progress program in the last quarter of fiscal year 1993. Due to the limited number of large U.S.-flag ships available to meet cargo preference requirements, this competition required Sri Lanka to split its wheat shipments among smaller U.S.-flag ships at higher rates.

Applying cargo preference to U.S. food aid programs can also adversely affect their operation by precluding recipient countries from purchasing commodities at the lowest available price or requiring a purchase of a different variety of commodity than originally planned. This situation occurs when no U.S.-flag ships are available at the ports where these commodities are located, or when those U.S.-flag ships available are not appropriate to carry the commodity. This can force commodity purchasing decisions to be driven by the availability of U.S.-flag ships, not the availability of the lowest cost or most desired commodity.

For example, for a recent P.L. 480 title I wheat purchase for Tunisia, Tunisia was unable to take advantage of the four lowest wheat offers because no U.S.-flag ships were available at the times when and locations where the wheat was available. To comply with cargo preference requirements, Tunisia was required to obtain more expensive wheat (over $4 more per metric ton) that was available where U.S.-flag ships were also available.

In addition, several P.L. 480 title I countries have been interested in obtaining western white wheat that is available from the West Coast of the United States. However, the availability of U.S.-flag ships on the West Coast is limited because West Coast cargos tend to be infrequent, and West Coast port charges are much higher than those in the Gulf of Mexico. This has forced the countries to obtain a different variety of wheat available in the Gulf of Mexico where U.S.-flag ships are available. Due to the limited availability of U.S.-flag ships, as well as on the overall difficulty that title I recipient countries experience with cargo preference requirements, USDA said that cargo preference requirements interfere with its ability to develop future commercial markets for U.S. agricultural products. This mission is one of the primary objectives of the title I program.

CERTAIN USDA AND AID MANAGEMENT PRACTICES MAY AFFECT FOOD AID TRANSPORTATION COSTS

Finally, several of the practices USDA and AID used to manage food aid transportation may affect transportation costs. One of these practices, used by both USDA and AID, was to require shipowners to accept contract terms that dictate that they arrange and pay for services that are typically the responsibility of the commodity supplier or buyer in commercial sales. These terms may increase transportation costs because they place additional costs and risks on the shipowners. These are then passed back to the food aid programs through higher shipping rates and may increase the total cost to transport U.S. food aid.

For example, for most landlocked countries, USDA and AID require shipowners to arrange and pay for transporting the commodity from the discharge port to its final destination. USDA and AID may also require shipowners to arrange and pay for any fumigation services required at the discharge port. USDA and AID choose to provide these services through the shipowners to provide additional financial assistance to these needy countries. Since shipowners must estimate the cost of providing these services before they are delivered and are paid based on their estimates, it is uncertain whether USDA and AID are paying more or less than the actual costs of providing these services.

MARAD officials, and the U.S. shipowners we talked to who carry food aid preference cargos, said they believe that adopting terms that are more closely aligned with terms used for transporting private sales of commodities would reduce the cost of transporting food aid. They said that using these types of terms would eliminate items for which actual costs are difficult to project and that increase shipping rates. As shipowners may overestimate the costs of providing these additional services--to cover their uncertainties about anticipated actual cost--providing the services through another means, such as reimbursing recipient countries for actual costs, could result in lower expenditures on the part of the government.

USDA and AID agreed that their contract terms added to food aid transportation costs, but they did not agree that adopting contract terms that are more consistent with those used for similar commercial cargos would reduce their food aid transportation costs. This is because items currently paid through shipping rates would still have to be paid through the food aid programs. While this may be true, given that estimated costs are used to pay the shipowners for these services, we believe that a potential may exist for some savings in food aid transportation costs by adopting transportation terms that are more consistent with those used for commercial transport of private commodity sales.

Another practice that increases food aid transportation costs is the concentration of food aid shipments into the last half of the year instead of spacing them more evenly throughout the year. For example, in 1992, 94 percent of the food aid tonnage under the P.L. 480 title I program was shipped between July and December. And for 1993, 73 percent of the food aid tonnage was shipped between July and December. This concentration of food aid shipments caused increased demand for the limited number of U.S.-flag ships available. This situation has resulted in higher U.S.-flag shipping rates as higher cost U.S.-flag ships have been used to meet the increased demand to transport food aid preference cargos.

USDA and AID officials said there is no conclusive reason why the majority of food aid tonnage is shipped in the last half of the calendar year. However, one reason they cited is that recipient countries delay signing their annual agreements with the U.S. government. Both USDA and AID officials said that while signing agreements earlier could result in some countries purchasing and shipping their food aid purchases sooner, there are other factors unrelated to the signing of agreements that could also affect the timing of shipments. These factors include the amount of commodity storage available in the country and the availability of alternative food sources. USDA and AID officials explained that they are reluctant to place pressure on recipient countries to sign agreements earlier and request food aid sooner because they believe it may discourage the countries from participating in the programs. Nevertheless, USDA said it will try to obtain the early signing of agreements and prompt purchasing of food aid commodities.

CONCLUSIONS

The application of cargo preference requirements to food aid programs does not significantly contribute to helping achieve the objectives of the Merchant Marine Act of 1936. The U.S.-flag ships that are most dependent on food aid preference cargos are not currently viewed by DOD as militarily useful. In addition, food aid preference cargos do not significantly contribute to ensuring that U.S.-flag ships carry a substantial portion of either U.S. domestic or foreign waterborne commerce.

Furthermore, the U.S.-flag ships that DOD currently views as militarily useful--those that provide liner service--are either supported by Jones Act trade or are largely dependent on operating subsidies, which are expiring, to successfully compete for foreign commercial cargos because their costs are substantially greater than their foreign competitors. In addition, U.S. shipowners are discouraged from taking the necessary steps to reduce their costs by investing in new ships because of the 3-year waiting period currently imposed on cargo preference eligibility for foreign-built U.S.-flag ships, and because it is possible to successfully operate inefficient ships in the food aid cargo preference trade. Thus, applying cargo preference laws to food aid programs makes it possible for U.S. shipowners to maintain inefficient and commercially noncompetitive U.S.-flag ships that do not significantly contribute to the ability of the U.S. merchant marine to carry foreign commerce other than food aid, and which increase the cost to transport U.S. food aid preference cargos.

In addition, the higher cost of using U.S.-flag ships, and their limited availability, can adversely affect the operation of U.S. food aid programs by (1) reducing the amount of commodity that can be purchased and (2) prohibiting a country from purchasing the lowest priced, or the variety of, commodity desired. This situation occurs because commodity purchasing decisions are driven more by the availability of U.S.-flag ships than the availability of commodities.

Food aid transportation costs can also be affected by USDA and AID management practices. USDA and AID have been using ocean transportation contract terms as a means to provide additional services to recipient countries. The shipowners must comply with these terms even though they may result in higher shipping rates. We believe that some potential may exist for savings in food aid transportation costs by removing the uncertainty associated with estimating the costs of these services by adopting terms more consistent with those used for similar commercial cargos.

Food aid transportation costs are also increased because a majority of food aid shipments occur in the last half of the calendar year instead of being more evenly spaced throughout the year. While there is no conclusive reason why this happens, it is clear that food aid transportation costs could be reduced if food aid shipments were more evenly spaced throughout the year.

MATTERS FOR CONGRESSIONAL CONSIDERATION AND RECOMMENDATIONS

In the report that we issued today, we included matters for congressional consideration regarding the continuance of cargo preference requirements for food aid programs. We also included recommendations to heads of agencies which, if implemented, should help reduce the transportation costs of food aid cargos.

Regarding the matters for congressional consideration, if Congress continues to support the objectives for which cargo preference is applied to food aid programs and is willing to continue to devote resources to that end, Congress may wish to consider a more efficient alternative for achieving those objectives. For example, a program like the current operating subsidy program, which will expire by 1998, could be used to support those ships, and their crews, that DOD finds militarily useful and that could also successfully compete for U.S. foreign commercial cargos.

If Congress decides to continue to apply cargo preference laws to food aid programs, it may wish to consider giving U.S. shipowners incentives to invest in more efficient ships in order to reduce food aid transportation costs, such as allowing new, foreign-built, U.S.-flag ships to immediately carry food aid preference cargos.

Regarding our recommendations, if Congress chooses to continue the application of cargo preference laws to food aid programs and acts to permit U.S.-flag foreign-built ships to immediately carry food aid preference cargos, we recommend the following:

-- The Secretary of Transportation should instruct the Administrator of the Maritime Administration to promote the efficiency of the ships that carry food aid preference cargos. One way this can be done is by changing the method of calculating guideline rates so that "average" operating costs for all similar-sized ships, instead of "actual" operating costs for each individual ship, are considered.

-- The Secretary of Agriculture and the Administrator of the Agency for International Development should take the following steps because of their potential to reduce food aid transportation costs:

-- experiment with the use of contract terms for the transportation of food aid cargos that are more consistent with contract terms used for similar commercial cargos to determine whether their use would reduce food aid transportation costs, and

-- encourage recipient countries to more evenly space their food aid shipments throughout the year.

While implementing these changes will not help the application of cargo preference laws to food aid programs achieve its intended objective, it should reduce food aid transportation costs.

- - - - -

Mr. Chairman and Members of the Subcommittee, this concludes my prepared statement. I will be happy to try to answer any questions you or other Members of the Subcommittee may have.

(280110)

Ordering Information

The first copy of each GAO report and testimony is free. Additional copies are $2 each. Orders should be sent to the following address, accompanied by a check or money order made out to the Superintendent of Documents, when necessary. Orders for 100 or more copies to be mailed to a single address are discounted 25 percent.

Orders by mail:

U.S. General Accounting Office
P.O. Box 6015
Gaithersburg, MD 20884-6015

or visit:

Room 1100
700 4th St. NW (corner of 4th and G Sts. NW)
U.S. General Accounting Office
Washington, DC

Orders may also be placed by calling (202) 512-6000 or by using fax number (301) 258-4066.

Each day, GAO issues a list of newly available reports and testimony. To receive facsimile copies of the daily list or any list from the past 30 days, please call (301) 258-4097 using a touchtone phone. A recorded menu will provide information on how to obtain these lists.

TOM BARLOW
1ST DISTRICT, KENTUCKY

1533 LONGWORTH BUILDING
WASHINGTON, D.C. 20515
(202) 225-3115
FAX: (202) 225-2169

Congress of the United States
House of Representatives
Washington, DC 20515

COMMITTEES:
AGRICULTURE
GENERAL FARM COMMODITIES
FOREIGN AGRICULTURE AND HUNGER
ENVIRONMENT, CREDIT AND RURAL DEVELOPMENT
MERCHANT MARINE AND FISHERIES
COAST GUARD AND NAVIGATION
NATURAL RESOURCES
ENERGY AND MINERAL RESOURCES

October 7, 1994

The Honorable Timothy J. Penny
Chairman, Subcommittee on
Foreign Agriculture & Hunger
U.S. House of Representatives
1336 Longworth HOB
Washington, DC 20515

Dear Chairman Penny:

I would appreciate it if you could include the enclosed materials in the record of the September 29, 1994, Subcommittee on Foreign Agriculture and Hunger hearing concerning cargo preference and food aid.

Thank you very much for your assistance.

Sincerely,

Tom Barlow

TOM BARLOW
Member of Congress

TJB/jca

THE JOURNAL of COMMERCE,

January 22, 1991

APL-Hired Ship's Crew Refuses Gulf Orders

By ALLEN R. WASTLER
Journal of Commerce Staff

The crew of a foreign-flag feeder vessel chartered by American President Lines Ltd. refused to enter the Persian Gulf last week, prompting the carrier to send in one of its U.S.-flag vessels instead, The Journal of Commerce learned.

The Eagle Nova, a feedership capable of carrying up to 400 20-foot containers, was scheduled to

American President Lines wants blanket authority from the government to shift its feederships in Mideast. Story, Page 8B.

deliver military goods to Dammam — a Saudi Arabian port in the Persian Gulf early last week.

The officers and crew of the vessel refused to enter the gulf because of the threat of war in the region, according to labor and industry sources.

"It's another instance of foreign-flag vessels with foreign-flag crews supposedly under the control of the government but not being under control," said Gunnar Lundeberg, president of the Sailors Union of the Pacific.

The President Buchanan, a U.S.-flag APL liner ship, was ordered to deliver the needed supplies to Dammam, not one of its regularly scheduled calls.

The American crew was issued gas masks and told to watch for 19 mines believed to be adrift in the area, according to a ship telex.

The Eagle Nova is registered in Cyprus. It is one of three small feeder vessels APL chartered to run military supplies to Saudi Arabia from Fujairah, a United Arab Emirates port situated along the Gulf of Oman.

APL declined to comment on the incident.

JOURNAL of COMMERCE,

January 24, 1991

Indian Air, Ship Crews Refuse Duty In the Gulf

By N. VASUKI RAO
And JANET PORTER
Journal of Commerce Staff

NEW DELHI, India — The government is facing a mutiny of sorts, with crews of Shipping Corp. of India and pilots of Air India, both state owned, refusing to take ships and aircraft to the Persian Gulf war region.

But international ship-management companies, which employ large numbers of Indian crewmembers, are faring better. Very few seafarers from India or anywhere else are refusing to sail into the gulf, according to several major ship-management concerns interviewed this week. The companies operate ships under several flags for a multitude of owners throughout the world.

Officials in India ordered all Indian ships to quit the Persian Gulf area ahead of the Jan. 15 U.N. deadline for use of force against Iraq. Air India simultaneously suspended its flights to the gulf.

This week, the government reversed itself and ordered Shipping Corp. of India to resume loading oil from Iran. After an outcry from Indians stranded in the region, Air India management decided to resume some services.

In both cases, however, crews are refusing to go to the area.

Air India Wednesday flew to Abu Dhabi, Dubai and Muscat, but its aircraft were piloted by executives. The first two are United Arab Emirates ports along the Persian Gulf, while Muscat is located well outside the Persian Gulf, in Oman.

Transport Minister Manubhai Kotadia appealed Wednesday to crews and officers of the shipping line to "rise to the occasion" and bring much-needed crude oil to the country. The bulk of India's oil is transported by the carrier and comes from the Middle East.

Mr. Kotadia reminded the crew that during the long Iran-Iraq war they had faced great risk and still brought oil from the region.

The shipping line hopes four tankers will load crude from Fujairah, on the coast of the United Arab Emirates in the Gulf of Oman outside the war zone. Two are due in port today, with the others arriving by the weekend.

Officials of the carrier acknowledge they are concerned about the safety of its crews. They intend to provide gas masks for them.

Crews will also be getting a 200% bonus for going into the war zone and insurance cover ranging from 300,000 to 500,000 rupees (US$17,000 to US$28,500).

As reported in The Journal of Commerce earlier this week, the crew of the Cypriot-flag Eagle Nova recently refused to enter the gulf to deliver U.S. military cargoes to Dammam, Saudi Arabia. Officials of leading ship-management companies report no similar experiences.

In London, Wescol International Marine Services said it has flown gas masks and protective clothing out to the Middle East and is paying seafarers a bonus of 200%. So far, none has refused to remain on a ship heading for Kharg Island, Ras Tanura or other Gulf destinations, according to John Wickham, chief operating officer for the Wescol group of companies.

At Hanseatic Shipping Co. in Cyprus, only two seafarers have so far refused to go to Ras Tanura in recent weeks, said Capt. Joachim Meyer, chairman and managing director. None of the crew on a gas carrier now loading in the gulf had asked to be replaced, he said Tuesday.

Finding replacement crews is not yet a problem, according to Mr. Meyer, although he admits the cancellation of airline services to the Middle East could create a problem if crews need to be flown out to join ships about to enter the gulf.

All operators insist that no crewmember would ever be forced to sail into a dangerous area against his wishes, and most are paying bonuses of between 100% and 200% of basic wages. But having continued trading throughout the Iran-Iraq war, most merchant seamen serving on tankers have plenty of experience operating in dangerous waters, said Mr. Wickham.

THE JOURNAL of COMMERCE, January 25, 1991

Some Ship Crews Resist Voyages Near War Zone

Journal of Commerce Special

Crewmembers of India's state-owned Shipping Corp. of India continued to resist sailing to Iranian oil loading points Thursday. Air India's regular pilots reversed their earlier stand and agreed to fly to the region.

Separately, Brazil's state-owned oil company reportedly fired the captain and threatened to replace the crew of a tanker that refused to sail into the Persian Gulf.

The oil company, Petrobras, flew a new captain from Rio de Janeiro to Muscat to assume command of the tanker Enrique Dias.

Capt. Jorge Teodoro da Silva was fired Wednesday, a day after he and the crew of the Enrique Diaz reportedly refused to head from the Port of Muscat in the Gulf of Oman to a United Arab Emirates port in the Persian Gulf to pick up a shipment of crude oil.

Reports from Rio de Janeiro said the captain had asked the crew if they wanted to enter the Persian Gulf and all 32 crewmembers voted no.

Maximino da Fonseca, Petrobras transport director, reportedly said the entire crew would be replaced if it disobeyed orders a second time.

In New Delhi, Shipping Corp. of India said two tankers reached the United Arab Emirates' port of Fujairah to load oil, with two others were on the way.

A bulk carrier also is proceeding

> In the air, pilots belonging to the Indian Pilots' Guild reversed their stance and agreed Wednesday night to undertake daytime flights to stations in the Emirates and Oman.

toward the Saudi port of Jeddah, a company statement said.

The announcement made no mention of any tanker moving to Iranian loading points, which is what the Indian government wants.

The Indian ship line said it is confident the "crucial supply of crude oil to Indian refineries would be maintained with the courage and patriotism of its officers and crew."

There are suggestions that other vessels may be chartered if more crude than already contracted for is to be lifted.

In the air, meanwhile, pilots belonging to the Indian Pilots' Guild agreed Wednesday night to undertake daytime flights to stations in the Emirates and Oman. Their decision followed talks with the civil aviation ministry.

When they initially refused to undertake these flights, operations were resumed by Air India executives.

WALL STREET JOURNAL, January 30, 1991

Foreign Crews' Fears Slow Some U.S. Gulf Shipments

By a WALL STREET JOURNAL Staff Reporter

NEW YORK—Some foreign crews of ships carrying U.S. military cargo are refusing to sail into the Persian Gulf, causing shippers to turn increasingly to vessels manned by U.S. citizens.

Although the disruption caused to military shipments has been minimal so far, U.S. shipping executives worry that foreign crews could become less reliable if commercial shipping in the Gulf comes under attack.

So far, industry executives count a half-dozen examples of delays to U.S. military cargo in the past two weeks caused by foreign crews' unwillingness to sail to northern Persian Gulf ports. Industry officials say the incidents highlight the danger of the U.S.'s growing reliance on foreign shipping as its own merchant fleet continues to shrink.

American President Cos., Oakland, Calif., said it recently directed one of its U.S. flag ships to Dammam, Saudi Arabia, when the crew of the Eagle Nova, a foreign-flag vessel the line had chartered, refused to take the cargo. George Hayashi, president of American President's shipping subsidiary, said the German master of the Eagle Nova "didn't want to endanger the lives of the crew and be an object of the shooting."

Lykes Bros. Steamship Co., New Orleans, said foreign crews on several occasions have unloaded U.S. military cargo bound for Saudi Arabia at ports outside the Persian Gulf, delaying the shipments by a couple of days.

THE JOURNAL of COMMERCE,
February 1, 1991

Crew Jumps Bangladeshi Ship Bound For Gulf

By BRUCE VAIL
Journal of Commerce Staff

WASHINGTON — Twenty-seven Moslem crewmembers of a Bangladeshi cargo ship deserted their vessel in Oakland, Calif., over the last two weeks, apparently in fear of being sent to the Persian Gulf war zone with U.S. military cargo aboard.

The desertions helped prompt the Pentagon to cancel its contract with the vessel owner, Bangladesh Shipping Corp., Chittagong, according to an executive of the local agent, General Steamship Corp.

Ashley Smith, operations manager for the San Francisco-based company, said the Immigration and Naturalization Service had been notified of the presence of illegal aliens but none of the fugitives had been apprehended as of early Thursday.

Crewmembers of the freighter Banglar Mamata were in the process of lining the cargo holds with plywood prior to loading military cargo hold when they jumped ship, Mr. Smith said.

He added he was unaware of the precise nature of the cargo or the proposed destination but he "presumed" it was ammunition bound for the gulf.

"They (the military) don't even tell the captain where he is going. They just tell him to sail in a certain direction and wait for further instructions," Mr. Smith said.

"The cargo manifest just lists plywood crates with numbers on them. They don't tell the captain what's in them . . . the charter terms are such we presume it's ammunition," he added.

Marge Holtz, director of legislative and public affairs for the Navy's Military Sealift Command, said the agency had chartered the ship Dec. 21, 1990, and had canceled the charter on Jan. 21. The reason for cancellation was "non-performance," she said.

Anwurel Haque, owner's representative for Bangladesh Shipping Corp. in New York, said "Naturally, civilian crew are scared to go the Persian Gulf. The war seems much more terrible from far away. We may know the fighting is confined to Kuwait and Iraq but all they see is there is war in the gulf."

Mr. Haque said 24 crewmembers and three officers had left the ship. "We don't have anyone on it anymore."

The owners representative added that the ship is currently unseaworthy and all plans to enter the gulf have been canceled.

THE JOURNAL of COMMERCE,
February 1, 1991

Sea-Land Experiences Problems With Crews of Foreign-Flag Ships

By BRUCE VAIL
Journal of Commerce Staff

WASHINGTON — An official of the largest container shipping company in the United States confirmed this week that it has encountered problems in getting foreign nationals to crew cargo ships entering the Persian Gulf war zone.

Sea-Land Service Inc., Edison, N.J., experienced an incident in January when the crew of a foreign-flag ship carrying military cargo balked at entering the gulf just as hostilities became imminent, a Washington-based executive of the company said this week.

The incident "was not significant, it wasn't a big deal," said Jack Helton, Sea-Land's vice president for government services.

Military cargo scheduled to be carried on the ship was not delayed, he said, and was loaded aboard a second foreign-flag ship the company employs in voyages between the Spanish Port of Algeciras and the gulf. The military cargo is carried on giant U.S.-flag ships between the United States and Algeciras.

The incident preceded by "a few days" a similar episode involving American President Lines Ltd., Oakland, Calif.

On Feb. 12, the German officers and Burmese crewmembers of the small containership Eagle Nova refused to enter the gulf, reportedly for fear of becoming a target in the war between the United States and Iraq.

American President Lines subsequently transferred the Eagle Nova's cargo to a U.S.-flag containership that delivered the goods to the Saudi Port of Dammam, a company spokesman said last week.

Outside sources said one problem encountered by Sea-Land involved Japanese officers aboard a containership that normally travels from Europe to the Persian Gulf and Asia. This service, operated by Sea-Land in conjunction with Swiss shipping company Norasia Ltd., uses no U.S.-flag vessels.

According to reliable sources, the Japanese officers aboard one vessel refused to take the ship into the gulf acting under a directive from the All Japan Seamens Union, a labor group that has also instructed Japanese tanker officers to avoid the gulf.

THE WASHINGTON POST
THURSDAY, FEBRUARY 7, 1991 D.C. 9

JACK ANDERSON and DALE VAN ATTA

Foreign Merchant Crew Balks at Gulf

President Bush has crafted a fragile alliance in the war against Iraq—governments around the world who have pledged money, arms and moral support. But that doesn't mean their citizens have to go along with it, including the crews of commercial foreign ships hired to carry supplies to the Persian Gulf.

In a recent incident, one crew of a German ship refused to sail into the gulf and the shipping company had to send in a U.S.-flagged ship to finish the job.

The U.S. Military Sealift Command, which oversees the naval supply operation, has played down the incident. But some on Capitol Hill aren't convinced and fear that the war could be hampered by more skittish delivery crews who refuse to sail into a combat zone.

The U.S. government uses commercial ships to carry everything from tanks to toilet paper for troops stationed overseas. The sailors belong to the private Merchant Marine. Since the American commercial fleet has dwindled substantially in recent years, the heavy shipping demands of a war force the Pentagon to sometimes rely on foreign ships.

And it was a foreign ship that balked about going into the gulf. That ship, the Eagle Nova, reportedly is owned by a German company. American President Lines of Oakland, Calif., hired the Eagle Nova to carry cargo from the United Arab Emirates to India, but at the last minute diverted the cargo to supply U.S. forces in the gulf.

When the crew refused to go, American President Lines sent a replacement ship under U.S. flag.

The House Merchant Marine and Fisheries Committee, chaired by Rep. Walter B. Jones (D-N.C.), had been keeping track of the Merchant Marine service during the Persian Gulf crisis. Our associate Scott Sleek has learned that the committee is now scrutinizing the Eagle Nova incident.

A sizable percentage of the gulf-bound cargo, including weapons and ammunition, is being shipped on foreign vessels. Iraq has a negligible navy, but can make life dangerous in the gulf with mines and missiles. Foreign crews may not feel enough loyalty to the anti-Iraq alliance to risk their necks supplying the troops.

Sealift capacity has been a worry from the onset of the gulf crisis. When Bush first sent troops to Saudi Arabia in August, the deployment was hampered by mechanical problems with the merchant vessels. Even if all the ships were in perfect working order, which they aren't, the United States still would not have enough American commercial ships to meet the demands of half a million troops sent half a world away.

In the late 1980s a special commission issued a series of reports warning about the decline in war readiness of the Merchant Marine fleet. "The deteriorated condition of America's maritime industries continues to present a growing danger to the national security," one of the reports concluded.

That leaves the Pentagon dependent on foreign ships and crews that may have no allegiance to the allied cause.

THE JOURNAL of COMMERCE,
February 11, 1991

Sealift Officer Hails, Frets About Foreign Vessel Charters

By WILLIAM DIBENEDETTO
Journal of Commerce Staff

WASHINGTON — Vice Adm. Paul D. Butcher said chartering foreign vessels to help supply troops in the Persian Gulf is a double-edged sword: the strategy has worked well but does nothing to enhance the U.S. merchant fleet.

Several factors have combined to make the $2.2 billion sea and air supply operation successful, said the deputy commander in chief of the U.S. Transportation Command in a phone interview last week from his Scott Air Force Base, Ill., headquarters.

They include the luxury of time — "We had 161 days to move the equipment" — and the "universal support" of the rest of the world, he said.

That support enabled the Navy to charter 91 foreign vessels that carried about 44% of everything the United States has shipped to the region. Another 81 vessels of the 213 used came from the Military Sealift Command or the reserve fleet; the Navy was able to charter only 41 U.S.-flag commercial dry cargo ships and tankers.

Without foreign charters and without the $7 billion invested in the Navy's sealift reserve fleet during the Reagan administration years, deadlines would not have been met and "we would have seen the real shortfall of the U.S. merchant marine," he said.

"I'm concerned about that reliance on foreign charters both in peace and war," the admiral continued.

"The country as a nation has to decide what we are going to do about the lack of strategic sealift; we can't depend on always having favorable factors for sealift," including readily available foreign charters, he added.

President Bush's budget for fiscal 1992 which begins Oct. 1 proposes $900 million for a new fast sealift vessel construction/acquisition program.

That money, however, is a carry-over of funds approved by Congress for the current fiscal year.

The budget message revealed little on how those dollars should be spent to beef up the quick response supply fleet, which currently totals eight aging vessels. It said that $30 million should go to buy a petroleum product tanker.

"Sealift was critical and continues to be critical," said John Stocker, president of the Shipbuilders Council of America.

Most of the total sealift to the gulf was hauled through the Navy's "own in-house capability," he said.

Mr. Stocker said the fast sealift program ought to focus on "military solutions to the sealift problem," like building more roll-on, roll-off vessels.

With the overall Naval shipbuilding program being reduced, "many yards will look at sealift with substantial interest," he added.

Rob Quartel, a member of the Federal Maritime Commission, recently said that ideas like a program to build commercially viable vessels that are useful for defense purposes "are really useless; you can't build ships that are both militarily and commercially useful."

The "real questions are about the labor pool," he said, and whether there should be a merchant marine reserve.

Adm. Butcher asserted that job protection, a feature of armed forces reserve programs, should be applied to merchant mariners who leave their regular jobs to work on reserve vessels.

"What should come out of this is to find a way to enhance the merchant fleet, and that would include a merchant marine reserve. Ten years from now, most mariners will be retired or dead. If you don't have an industry with a future, who's going to do it?" he asked.

During the interview, Adm. Butcher also said:

• There have been no big adjustments in the air and sea supply lines since the war started.

• Sealift is slower, but cheaper: From Aug. 2 to Dec. 31, it cost $906 million. As of last week 88%, or 5.1 million tons of the cargo had gone to the Persian Gulf by sea.

• Airlift is faster, but more expensive: It cost $1.3 billion for the military airlift as of Dec. 31. By last week, 448,000 people and 441,000 tons of cargo were flown to the region.

• The supply exercise has stressed a need within Transcom to "manipulate data faster."

STATEMENT OF PHILIP J. SHAPIRO
OF
LIBERTY MARITIME CORPORATION

BEFORE THE SUBCOMMITTEE ON
FOREIGN AGRICULTURE AND HUNGER

HOUSE COMMITTEE ON AGRICULTURE

SEPTEMBER 29, 1994

My name is Philip Shapiro and I am president and chief executive officer of Liberty Maritime Corporation. I appreciate the opportunity to submit this statement for the record.

Liberty operates five U.S.-flag dry bulk carriers as well as a tanker. Our dry bulk carriers, constructed in Korea during the mid-1980s, are among the most modern and efficient in the U.S.-flag fleet. While Liberty competes vigorously in international markets, our primary market is the transportation of U.S. government food aid under the P.L. 480 and section 416 programs. Under the cargo preference laws, not less than 75 percent of food aid cargoes must be transported on U.S.-flag vessels, if available at fair and reasonable rates. USDA is responsible for two-thirds of this cost, with DOT responsible for the remainder.

Cargo Preference and Food Aid

Before turning to the GAO report, I think it is important that we step back and view cargo preference and food aid in their proper context. While we all recognize that food aid programs serve a vital humanitarian and foreign policy purpose, all too often we forget that these programs, through Buy-American requirements for both commodities and shipping, create jobs and economic opportunity here in the United States for farmers, food processors, ports, maritime, and other important sectors of our economy.

While our commodities and shipping sometimes may cost more than those of our foreign competitors, which all too often are heavily subsidized and/or virtually exempt from burdensome tax and regulatory requirements, most Americans would be outraged if our government decided to provide food aid in the form of heavily subsidized foreign-grown commodities shipped on untaxed, substandard, subsidized foreign-flag vessels. They would easily recognize the fundamental unfairness of the federal government imposing costly tax and regulatory requirements on American maritime and agriculture while, at the same time, purchasing foreign-grown commodities and foreign-flag shipping just because they may be less expensive.

Such a course of action would be not only bad politics, but also bad budget policy, since the tax revenues resulting from the economic activity here at home generated by the food aid programs likely offset the relatively small additional budgetary expense of buying and shipping American. GAO, for example, estimates that the $600 million in food aid cargo preference spending in FY 1991, 1992, and 1993 generated $1.5 billion in revenue for U.S.-flag operators during this period (page 48). Undoubtedly, much of this $1.5 billion was returned to the U.S. government as tax revenue, either directly or indirectly. Therefore, the effect of cargo preference on federal outlays is minimal and may even be positive. Unfortunately, GAO made no effort to determine the budgetary impact of preference net of recycled tax revenues. Proponents of USDA's various export promotion programs -- credits, credit guarantees, export enhancement, market promotion, and food aid -- can and do make the same point to support their programs.

Both the agriculture and maritime sectors have come to recognize that preserving the food aid programs is an important national goal and are working, along with the Coalition for Food Aid and its private voluntary organization (PVO) members such as CARE, Catholic Relief Services, and Africare, to ensure the continuation of these programs in both the appropriations process and in the upcoming farm bill. As this committee is aware, food aid spending has declined from $2.3 billion in FY 1993 to $1.6 billion in FY 1994 to $1.3 billion in FY 1995 (P.L. 480 and Section 416). Food aid tonnage has dropped correspondingly, from nearly eight million metric tons in FY 1993 to less than five million metric tons in FY 1995.

The drastic cuts in the food aid budget dwarf the cost of cargo preference, which, as noted, was only about $600 million over a three-year period, with USDA responsible for $400 million and DOT responsible for $200 million. During this period, the U.S. donated about $5.1 billion in American commodities -- almost twelve times USDA's cargo preference costs. Put another way, even with the elimination of food aid cargo preference and the shift of all of USDA's cargo preference funding to food aid commodity purchases -- a highly unlikely event given today's budgetary austerity -- USDA's food aid spending would increase by only 7.8 percent (page 49).

It also bears emphasis that cargo preference affects only a small portion of U.S. government-subsidized agricultural exports and an even smaller percentage of total agricultural exports. According to GAO, food aid constituted 18.1 percent of subsidized agricultural exports and 6.7 percent of total farm exports (pages 21-22), with the 25-percent foreign-flag shipping allowance reducing the impact of the U.S.-flag shipping requirement to 13.5 percent and 5.0 percent, respectively.

Given the minimal impact of cargo preference on agriculture and on the federal budget and the importance of food aid programs to both maritime and agriculture, it makes sense for (1) agriculture, maritime, and the PVOs to work together to preserve the food aid programs and fight budget cuts and (2) agriculture, maritime, MarAd, USDA, and AID to cooperate in finding more efficient ways to transport commodities aboard U.S.-flag vessels, as GAO recommends.

The GAO Report

Unfortunately, the GAO report will do little to advance these efforts, with the important exception of its recommendation that USDA and AID reform their ocean transportation procurement practices.

Ocean Transportation Procurement Reform. The one area where GAO is correct is in its recommendation that USDA and AID adopt common-sense commercial practices, which will result in lower ocean transportation costs.

According to USDA, the average rate for foreign-flag bulk carriers is $39 per ton and the average rate for U.S.-flag carriers is $69 per ton (page 38). GAO states that USDA and AID use "contract terms not typically required of shipowners for commercial cargoes" that increase transportation costs because "they place additional costs and risk on shipowners that are then passed back to the food aid programs through higher shipping rates" and that the concentration of food aid shipments in the last half of the year results in a high demand for shipping services and therefore increases costs (page 56).

GAO found several contract provisions used by USDA and AID that differ from standard commercial practice. First, full berth terms require the shipowner, rather than the receiver or charterer, to pay the cost of loading and discharge and to assume the risk of delay -- a requirement that MarAd found adds $15 per ton to the cost of a 50,000 ton cargo (page 57). In commercial practice, the receiver or charterer, either of which is in a better position than the shipowner to ensure quick and efficient loading and discharge, is responsible and must compensate the owner if there is a delay (demurrage) and receive compensation if loading and discharge are especially quick (despatch).

Second, instead of receiving 100-percent payment upon loading as is standard in the commercial practice, the shipowner receives 95-percent payment upon vessel arrival at its destination and 5-percent payment after the vessel completes discharge. USDA's claim that this practice is necessary to ensure cargo delivery is based on the ridiculous assumption that shipowners would risk their relationship with their primary customer. MarAd estimates that this practice adds up to $50,000 in unnecessary interest costs for a voyage (page 58).

Third, instead of inland transportation being the charterer's responsibility as is standard in commercial practice, the shipowner is often required to arrange it. Noting its amazement at USDA's and AID's practice, GAO states, "It seems reasonable to expect that some landlocked countries should be familiar with transporting goods inland, and thus should be able to make these arrangements at lower cost than shipowners who are less familiar with making these arrangements" (page 58).

Fourth, contrary to commercial practice, the shipowner is responsible for fumigation if the cargo is infested with pests, even though USDA inspectors certify vessels carrying food aid as pest-free prior to loading. MarAd estimates that the cost and delay resulting from fumigating can add $200,000 to voyage costs, which must be built into the shipping rate (page 59).

The cost savings resulting from the use of commercially-based charters and long-term procurement are readily apparent in GAO's discussion of practices used by Israel to arrange the purchase and shipment of U.S. commodities aboard U.S.-flag vessels pursuant to its "Side Letter" agreement with the U.S. GAO found that Israel charters U.S.-flag bulk carriers in the low $30 per ton range (page 76). In fact, the most recent rates for the Israeli trade are in the high $20 per ton range -- significantly less than the foreign-flag rates paid by USDA and less than half of what USDA pays for U.S.-flag bulk carriers. While it is true that the food aid programs cannot and should not be operated on a purely commercial basis, the Israeli experience indicates the huge potential savings resulting from even partial adoption of GAO's recommendations.

Vessels Militarily Useful. The crux of the report is GAO's misguided assumption that the principal reason for food aid cargo preference is to provide DOD with vessels it can charter for military shipments during wars and national emergencies and its conclusion that the program does not meet that objective because DOD chartered only liner vessels and roll-on/roll-off vessels during Operation Desert Shield/Desert Storm (page 32).

GAO is wrong. First, bulk carriers, tankers, and tug/barge units have costly military features demanded by DOD. Most of the vessels transporting bulk food aid cargoes were built under either MarAd's construction subsidy program or MarAd's build foreign program (46 U.S.C. app. §§ 1151, 1185). Under both programs, the vessels were required to be constructed with national defense features certified by the Department of the Navy. These include strengthened decks and holds to carry tanks and other heavy equipment, sophisticated communications equipment, prohibition of cast iron, washdown equipment for atomic, biological, and chemical warfare attacks, etc.

These Navy-requested features increased the construction cost of the vessels, were mostly borne by the shipowners under the construction subsidy program and wholly borne by the shipowners under the build foreign program, and are one of the reasons why U.S.-flag rates are higher than those of their foreign-flag-of-convenience competitors.

It is illogical for DOD -- only eight years later -- to switch and declare that the vessels are no longer militarily useful, when a national defense certification was a condition precedent for their construction and the vessels were required to have features useful only to DOD.

Second, Operation Desert Shield/Desert Storm is not a valid indicator of military usefulness. This conflict was unique in world history because there was a long mobilization period, virtually the entire world was united against Iraq, and Saudi Arabia provided DOD with perhaps the world's most modern deep-water, containerized port, combined with a virtually limitless source of inexpensive oil and oil products and state-of-the-art desalinization facilities. The next conflict may require the use of bulk carriers, tankers, and tug/barge units to carry grain, water, oil and other products, as well as the liner vessels that carry individual parcels. The definition of "militarily useful" can change radically in a different conflict under different conditions, where there could be vessel losses, such as in the Iran/Iraq war, that could raise the value of any type of vessel.

With this in mind, DOD cannot exclude all but the most ideal militarily useful vessels. During the Falklands War, for example, British oil and water tankers made a "vital contribution" to the war effort, during which Britain was "without allies or friendly bases near at hand" (Villar, Merchant Ships At War: The Falklands Experience 1984).

Third, DOD's focus on vessels it charters is too narrow. Bulk carriers, tankers, and tug/barge units are an important national security asset, even if DOD is not the user. Because foreign-flag-of-convenience shipping may not always be available to U.S. shippers, a base of U.S.-flag vessels is important to our national defense and our ability to move our goods in times of conflict. Similarly, foreign-flag vessels may not always be available to USDA and AID for transporting food aid to our friends and allies, if, for example, they were subject to a boycott.

Fourth, maintaining our food aid shipping capability has an important foreign policy justification. The people of countries receiving food aid are more likely to appreciate American generosity if they see our food arriving on vessels flying the American flag, particularly when the commodity is in bulk form and is off-loaded into a grain elevator.

Fifth, at a time when DOD is spending billions for a handful of reserve fast-sealift vessels that have no commercial utility, the roughly $200 million per year spent by USDA and DOT on food aid cargo preference is a bargain. As GAO recognizes, in 1991, 1992, and 1993, 144 privately-owned U.S.-flag liners, tankers, and bulk carriers and 71 tug/barges carried 17.1 million tons of food aid, with food aid providing most of the cargoes for some vessels and a significant amount for others. With respect to liner vessels, which even DOD regards as militarily useful and which are not as dependent on food aid as other types of vessels, food aid provides an essential margin of cargo. This is especially important given the unfortunate Senate delay in considering the President's maritime reform legislation (H.R. 4003/S. 1945), which would provide for annual payments to a limited number of U.S.-flag liner vessels.

Crews Militarily Useful. GAO mischaracterizes DOD's position on the military usefulness of crews who serve aboard vessels primarily transporting food aid cargoes. In response to GAO's inquiry, DOD stated, "Crews on all vessels are important because they are generally interchangeable, and provide support for manning of the Ready Reserve Fleet (RRF)" (page 104). As recognized by DOD, any skilled, experienced American seaman is useful during a war or national emergency. It would cost DOD untold millions of dollars to maintain a similar manpower base without cargo preference -- a fact not considered by GAO.

Operation Desert Shield/Desert Storm, as noted, provided almost ideal conditions for a massive military build-up (long lead time, international cooperation, state-of-the-art ports, etc.). Yet the demand for skilled American seafarers was so strong that retirees were called up to man both RRF vessels and commercial liner vessels chartered by DOD. Even worse, foreign seamen were hired to serve on U.S.-flag vessels because no active or retired American seamen were available. As pointed out by DOT, it is axiomatic that the loss of the U.S.-flag vessels now involved in food aid would worsen an already severe manpower shortage (page 142). DOD itself has expressed concern about the "reduced pool of mariners available to crew RRF ships when activated" (Statement of Ronald R. Fogelman, Commander in Chief, U.S. Transportation Command, House Committee on Merchant Marine and Fisheries, March 9, 1994).

Merchant Marine Supported. GAO opines that U.S.-flag bulk carriers and tug/barge units are virtually dependent on food aid preference cargoes and therefore contribute little to the ability of the U.S. to carry foreign commerce other than food aid.

GAO fails to recognize that the competitive problems faced by U.S.-flag vessels do not result from the cargo preference program. Indeed, the very existence of cargo preference reflects the longstanding congressional recognition that it is more expensive to operate as a U.S.-flag carrier than as a foreign-flag-of-convenience carrier because of an array of U.S. government-imposed tax, safety, environmental, health-benefit, and labor requirements. Federal and state income taxes, social security, Medicare, and unemployment alone double labor costs.

The foreign-flag-of-convenience vessels that carry the 25 percent of food aid not reserved for U.S.-flag vessels are manned not by nationals of our trade competitors, such as Germany and Japan, who receive salaries and benefits comparable to or even exceeding those received by American seamen, but instead by seamen from Bangladesh, the Philippines, and other third-world nations, who work for as little as $18 per day, with no benefits or tax obligations. Cargo preference is simply a decision by the U.S. government that it will help offset the cost of U.S.-flag operation by reserving a portion of U.S. government-generated cargoes to U.S.-flag vessels, if available at fair and reasonable rates.

Even though it will always be more expensive to operate under the U.S. flag than a flag-of-convenience, our merchant marine is becoming increasingly competitive. Cargo preference provides a base from which U.S.-flag vessels can expand their operations as their costs become similar to those of their foreign-flag competitors. The maritime reform legislation approved by the House and pending in the Senate addresses part of this problem by allowing new foreign-builds. Under MarAd's now-expired build-foreign program of the mid-1980s, seven new foreign-built vessels entered the cargo preference trade, which, as GAO found, cut the rate gap between U.S. and foreign vessels by half. In addition, legislation approved by the House on September 22 and pending in the Senate begins to address the problem of U.S. government regulation (Coast Guard) that is more burdensome than that of other nations (H.R. 4422). Under these initiatives, the U.S.-flag share of foreign commerce should increase significantly from the four-percent level it is today.

No Impact on Food Aid. GAO accepts without question the assertions of USDA and AID that food aid negatively affects food aid programs. GAO is wrong.

First, when Congress determines funding levels for food aid programs, it builds the additional cost of using U.S.-flag vessels into the funding. If cargo preference were eliminated, there is little chance that the amount budgeted for food aid would increase, because OMB would move to recover the savings.

Second, as pointed out above, the additional cost of using U.S.-flag vessels is only a small part of total food aid spending, and the elimination of cargo preference and the transfer of all USDA cargo preference spending would increase the total value of commodities shipped by only 7.8% (assuming OMB would allow it) (page 48). Given this context, it is obvious that AID's Sri Lanka and Mozambique anecdotes are misleading and are not at all representative of cargo preference's de minimis effect on food aid.

Third, GAO incorrectly gives credence to USDA's longstanding contention that commodity purchasing decisions and delivery times can be driven by the geographic availability of U.S.-flag vessels, rather than the geographic availability of the lowest priced or most desired commodity. USDA has been making this claim since at least 1955, when the House Merchant Marine and Fisheries Committee stated, "There was unanimity among all witnesses from both Government and industry on one point: there is no basis in fact for the charge that shipments to foreign countries of surplus commodities are being delayed by reason of the Cargo Preference Act" (H. Rep. No. 80, 84th Cong., 1st Sess. 18-19 (1955)). USDA is just as wrong now as it was then.

By law, U.S.-flag vessels must be used only if they are "available." If they are not available, foreign-flag shipping can be used. All USDA has to do is comply with the 75 percent U.S.-flag shipping requirement over the course of the year. Indeed, according to MarAd, USDA has fallen short of the 75 percent requirement because of supposed unavailability of U.S.-flag vessels. The situations cited by USDA and AID as evidence of the detrimental impact of cargo preference could have been avoided had they employed better management techniques.

Before casting aspersions on cargo preference, USDA should keep in mind the widely held view that the commodities we make available under the food aid programs often are more expensive than those available on the world market and often are not the ones desired by food aid recipients (Forbes, August 16, 1993, page 40; Journal of Commerce, page 9A, Oct. 4, 1993). U.S.-flag operators have no objection to USDA administering food aid programs to benefit American agriculture (consistent with the cargo preference laws), but are concerned when USDA makes unsubstantiated accusations about cargo preference that unfairly undermine support for the program.

Conclusion

Once again, I appreciate the opportunity to submit this statement for the record.

I hope that the maritime and agriculture communities can work together to preserve our nation's food aid programs. They are just as important here at home as they are in the world at large. If both industries step back, we will see that more unites us than divides us.

Thank you again for the opportunity to submit this statement for the record. I look forward to working with this subcommittee in the coming years.

September 29, 1994

The Honorable Timothy Penny
Chairman, Foreign Agriculture
and Hunger Subcommittee
Agriculture Committee
1336 Longworth House Office Building
Washington, D.C. 20515

Dear Mr. Chairman:

OMI Corp. ("OMI") respectfully requests that the following comments be incorporated in the record of the Subcommittee on Foreign Agriculture and Hunger regarding the report from the General Accounting Office entitled Cargo Preference Requirements; Objectives Not Met When Applied to Food Aid Programs.

On June 17, 1993, OMI presented testimony to the Subcommittee on the transportation of grain to Russia and other former Soviet republics, addressing primarily the concerns that had been raised about rates being charged by U.S-flag vessel owners for these shipments. This testimony presented to the Subcommittee specific rates charged by OMI, a comparison of OMI's rates to foreign flag vessel rates, and, in response to accusations of price gouging on the part of U.S.-flag vessel owners, evidence showing that OMI's rates were virtually identical to its rates of the previous year.

In that testimony, OMI identified a number of management problems in the cargo preference program that were directly responsible for the higher rates being charged for some preference shipments. OMI also noted that if the cargo preference program were managed in a more efficient manner using standard commercial terms, rates would not be as high and the cost of the program could be reduced.

OMI is gratified to see that the General Accounting Office report supports this position. GAO concludes: "... given that estimated costs are used to pay the shipowners for these services, we believe that a potential may exist for some savings in food aid transportation costs by adopting transportation terms that are more consistent with those used for commercial transport of private commodity sales." (Testimony, p.18.) GAO suggests USDA and AID experiment with more consistent commercial terms for services normally provided by a shipowner with the cost of any additional services reimbursed on an actual cost basis. (Testimony, p.24.)

OMI Corp. 90 Park Avenue, New York, NY 10016
(212) 986-1960 Fax: (212) 297-2100
Telex: RCA 224060 A/B OMI UR, TRT 177309 A/B OMI UT Cable: Bulktrans, NY

OMI fully supports the suggestion that a pilot program be designed and operated based on standard commercial practices. OMI is most willing to work with USDA, AID, and the Maritime Administration to design and institute such a program.

Providing efficient transportation of food aid cargoes is in the best interests of U.S.-flag vessel owners and the agricultural community. Both the food aid program and the cargo preference program support American jobs and American industries. Improving the efficiency and the effectiveness of the transportation portion of the program should be a priority for all who support these programs.

OMI looks forward to working with the agricultural community in this important endeavor.

Very truly yours,

Craig Stevenson

Craig Stevenson
Senior Vice President - Chartering

U.S.GOVERNMENT PRINTING OFFICE : 87-784

APEX
TP
9780160468070
76901945-69